TOXIC WASTE MINIMIZATION IN THE PRINTED CIRCUIT BOARD INDUSTRY

TOXIC WASTE MINIMIZATION IN THE PRINTED CIRCUIT BOARD INDUSTRY

by

T. Nunno, S. Palmer
M. Arienti, M. Breton

Alliance Technologies Corporation
Bedford, Massachusetts

NOYES DATA CORPORATION
Park Ridge, New Jersey, U.S.A.

Copyright © 1988 by Noyes Data Corporation
Library of Congress Catalog Card Number: 88-22630
ISBN: 0-8155-1183-3
ISSN: 0090-516X

Printed and bound in the United Kingdom

Transferred to Digital Printing, 2011

Library of Congress Cataloging-in-Publication Data

Toxic waste minimization in the printed circuit board industry / by T.
    Nunno . . . [et al.] .
          p. cm. -- (Pollution technology review, ISSN 0090-516X ; no.
    162)
        Bibliography: p.
        Includes index.
        ISBN 0-8155-1183-3 :
        1. Printed circuit industry--Waste disposal--Case studies.
    2. Printed circuit industry--Waste disposal--Evaluation.
    3. Semiconductor industry--Waste disposal--Case studies.
    4. Semiconductor industry--Waste disposal--Evaluation. 5. Hazardous
    waste treatment facilities--Case studies. 6. Hazardous waste
    treatment facilities--Evaluation. I. Nunno, T. II. Series.
    TD899.P69T69   1988
    621.381'74--dc19   88-22630
                                                   CIP

Foreword

This book presents information on waste minimization practices currently employed in the printed circuit board (PCB) and semiconductor manufacturing industries. Case studies conducted at six facilities evaluated the technical, environmental and cost impacts associated with the implementation of technologies for reducing the volume and toxicity of PCB metals-containing sludges and solvent wastes. The analyses of these data are the basis for demonstrating waste minimization technologies to reduce hazardous waste.

With the enactment of the Hazardous and Solid Waste Amendments in November 1984, Congress set forth a schedule for evaluating the land disposal restriction of various classes of hazardous wastes. A key issue identified in the evaluation of the waste bans is the availability of commercial treatment capacity to handle the wastes proposed for banning. Therefore, Congress also asked EPA to evaluate the potential for onsite waste minimization to reduce the quantity or toxicity of wastes being considered under the ban.

The electronics industry was initially judged as a good choice for individual case studies because it is a growth-oriented industry and ranks in the top 20 industries generating solvent wastes. The criteria for selecting case studies was further narrowed down to those facilities generating waste described by RCRA as waste treatment sludges from electroplating operations, and spent halogenated solvents or still bottoms from recovery of those solvents. These waste types were selected because they are two of the largest volume hazardous waste streams generated by the electronics industry.

The six case study assessments in the book use the results of analytical measurements to discuss the performance of each technology. In addition, measurements of process residuals and/or other discharges are presented. Finally, an assessment of the economics of each technology is also given to assist the cost evaluation of each technology.

Each facility investigated employs some practice that requires offsite disposal. Two of the case studies focus on the recovery of spent halogenated solvents, and the remaining four discuss the recovery or reduction of metal plating and etching process wastes.

The information in the book is from *Waste Minimization in the Printed Circuit Board Industry— Case Studies,* prepared by T. Nunno, S. Palmer, M. Arienti, and M. Breton of Alliance Technologies Corporation for the U.S. Environmental Protection Agency, January 1988.

The table of contents is organized in such a way as to serve as a subject index and provides easy access to the information contained in the book.

Advanced composition and production methods developed by Noyes Data Corporation are employed to bring this durably bound book to you in a minimum of time. Special techniques are used to close the gap between "manuscript" and "completed book." In order to keep the price of the book to a reasonable level, it has been partially reproduced by photo-offset directly from the original report and the cost saving passed on to the reader. Due to this method of publishing, certain portions of the book may be less legible than desired.

ACKNOWLEDGMENTS

The authors wish to extend their thanks and appreciation to Harry M. Freeman, the U.S. EPA technical project manager, and Lou H. Garcia, the U.S. EPA project officer and a peer reviewer of this report. Thanks are also extended to R.L. Stenburg of U.S. EPA and Mike Crawford of Metcalf and Eddy, Inc. for their peer review and comments. The authors would also like to acknowledge the contributions of Dr. James Patterson of the Illinois Institute of Technology for his assistance in formulating the scope of this project, and the printed circuit board industry representatives who participated in this study.

NOTICE

Contents and Subject Index

1. Introduction and Summary

BACKGROUND

With the enactment of the Hazardous and Solid Waste Amendments (HSWA) in November 1984, Congress set forth a schedule for evaluating the land disposal restriction of various classes of hazardous wastes including: (1) solvents; (2) metals and cyanides; (3) halogenated organics; (4) corrosives; and (5) dioxin wastes. A key issue identified in the evaluation of the waste bans is the availability of commercial treatment capacity to handle the wastes proposed for banning. Therefore, Congress also asked EPA to evaluate the potential for onsite waste minimization to reduce the quantity or toxicity of wastes being considered under the ban.

In an effort to identify successful waste minimization technologies, EPA's Office of Solid Waste (OSW) and Office of Research and Development (ORD) Hazardous Waste Engineering Research Laboratory (HWERL) set forth on research efforts aimed at assessing the viability of waste minimization as a means of reducing the quantities of land disposed hazardous waste. OSW's research focused on an exhaustive literature review identifying a broad spectrum of waste minimization technologies and their various applications. The primary emphasis of HWERL's work was on demonstrating the effectiveness of specific minimization technologies through case studies and process sampling.

WASTE MINIMIZATION CASE STUDY SELECTION

The case study development work was divided into two phases with Phase I involving:

- Waste category assessments;

- The identification of the data requirements and organization of the case studies; and

- The selection of specific sites/streams for use in the case studies.

The waste category assessments were a series of five reports aimed at identifying key industries that generate wastes which are being considered for restriction from land disposal. The five waste categories assessed included: (1) solvent wastes; (2) metals-containing wastes; (3) cyanide and reactive wastes; (4) halogenated organic nonsolvent wastes; and (5) corrosive wastes. The findings of these reports were used in conjunction with the findings of other aspects of the case study selection approach to help direct the final selections.

1

As part of the case study identification/selection process, the project team contacted trade associations and state agency representatives to solicit ideas and advice. As a result of these meetings, it was determined that case study selection should focus on a single industry or waste stream. The electronics industry was initially judged as a good choice because it is a growth-oriented industry and ranks in the top 20 industries generating solvent wastes.

The criteria for selecting case studies was further narrowed down to those facilities generating waste described by RCRA codes F006 or F001 and F002, which are respectively, waste treatment sludges from electroplating operations, and spent halogenated solvents or still bottoms from recovery of those solvents. These waste types were selected because they are two of the largest volume hazardous waste streams generated by the electronics industry, particularly by manufacturers of printed circuit boards and semiconductors.

Facilities which met the selection criteria were contacted to determine whether they practiced some form of onsite waste minimization or recycling. Preliminary site visits were scheduled for cooperating facilities after determining the willingness to participate. The purpose of the preliminary site visit was to evaluate the practicality of testing the waste minimization process, to determine its performance and to gather information necessary to conduct the testing.

During the case study selection process over 50 facilities were contacted by mailings or telephone to explain the case study program and determine their interest and anticipated level of cooperation. Based on the initial screening, 15 metals waste case studies and 12 resist strip solvent case studies were identified. Ten facilities were visited for pretest site visits to assess the facility's suitability for testing and further explain the intent and scope of the case study program. In the final section, six facilities were determined to be suitable to the scope of the program and willing to cooperate.

Part II of this study was devoted to testing waste minimization processes and developing the case study reports. During this phase of work under this program QA Project Plans (Test Plans) were prepared for the testing proposed at the six facilities selected in Phase I. Following approval of the Test Plan by the facility and EPA, testing was conducted. During the case study testing, process information was collected by the investigators or provided by the facility where appropriate. Mass throughput data and samples for analyses were collected according to the test plans. These data were occasionally supplemented by plant-supplied data where necessary to obtain a more representative picture of the long-term operation.

The case study assessments presented in this report discuss the results of analytical measurements used to discuss the performance of each technology. In addition, measurements of process residuals and/or other discharges are presented in the case studies. Finally, an assessment of the economics of each technology is also presented to assist the cost evaluation of each technology.

The remainder of this section presents a summary of each of the six case studies under review. Sections 2 and 3 present the conclusions of the project and recommendations for further research efforts. Section 4 presents pertinent production and waste management information on the electronics products manufacturing. Sections 5 through 10 present waste minimization case study results for facilities A through F, respectively. The Quality Assurance/Quality Control statistics for each case study are summarized and included as Appendix A.

PROJECT SUMMARY AND RESULTS

The purpose of this project was to evaluate the effectiveness of various waste minimization practices or technologies in the printed circuit board and semiconductor manufacturing industries. The most significant waste streams in these industries are waste halogenated solvents from photoresist stripping and developing operations (RCRA Waste Code F001-F003), and metal-bearing sludges (RCRA Waste Code F006) from the treatment of metal plating and etching rinsewaters. This project summary presents the findings of case studies conducted at five printed circuit board manufacturing facilities and one commercial treatment/recovery facility. Each facility investigated employs some practice that requires offsite disposal. Two of the case studies focus on the recovery of spent halogenated solvents, and the remaining four discuss the recovery or reduction of metal plating and etching process wastes. Table 1 summarizes characteristics of facilities investigated which range from small job shops to large integrated facilities.

TABLE 1.   SUMMARY OF FACILITIES TESTED UNDER WASTE MINIMIZATION
CASE STUDY PROGRAM

Facility (name)	Description	Wastes Treated/Reduced	Technology	Residuals
A	Treatment storage disposal facility handling electroplating baths, waste etchants, spills, etc. Capacity: 1,000 gph (24,000 gpd).	–Nickel plating baths –Copper plating baths –Cyanide	–Sodium hydroxide precipitation –Sodium borohydride reduction –Alkaline chlorination	Sludge product
B	Contract PC board manufacturing shop. Employees: 77 Production: 500,000 ft^2/yr Sales: $7 million/yr	–Cupric chloride etchant –Electroless plating rinses –Electroplating rinses	–Sodium borohydride reduction –Memtek ultrafiltration system	Sludge product
C	Computer manufacturer. Employees: 10,000	–Methyl chloroform resist developer –Freon resist developer	–Solvent distillation/fractionation recovery of resist developers.	Still bottoms
D	Electronic equipment mfgr. PC board manufacturing using the subtractive technique in the MacDermid process. Employees: 260	–1,1,1-Trichloroethane resist developer –1,1,1-Trichloroethane still bottoms	–2-stage solvent distillation –(1) DuPont RISTON SRS-120 solvent recovery still –(2) Recyclene Products, Inc. RX-25 still	Still bottoms
E	Computer manufacturer. PC board manufacturing using additive techniques. Employees: 600 Production: 600,000 ft^2/yr	–Acid copper plating bath	–Activated carbon regeneration of spent plating baths.	Spent activated carbon
F	PC board manufacturer. 2-sided single layer circuit boards. Production: 480,000 ft^2/yr	–Acid copper plating rinsewaters –Tin/lead plating rinsewaters	–Agmet Equipment Corp. electronic recovery units	Metal foil

Metal Plating Bath Waste Minimization Case Studies

Metal plating wastes generated from plating bath dumps, rinses, etching machines and scrubbing operations generate copper-, nickel-, tin-, and lead-contaminated wastes. Four of the six case studies investigated under this research project focus on the minimization of sludges generated primarily by copper plating and etchant baths and copper and tin/lead rinsewaters.

The common objectives of each of the technologies evaluated are: (1) minimization of metals sludges generated; (2) compliance with effluent guidelines or local discharge limitations; and (3) reduction in operating costs over other conventional alternatives. The following discussion briefly summarizes each case study, the nature of the minimization technology, the measurements data collected and the results obtained.

Facility A Case Study--

Description--Facility A is an offsite Treatment, Storage, and Disposal (TSDF) facility which processes concentrated dumps from the metal plating and printed circuit board industries, including alkaline etchants, acid plating baths, nitric acid rack strip baths, and electroless plating cyanide baths. The average total metals concentration in the incoming waste was reportedly 12 g/L (12,000 ppm). These waste streams are classified into the following four categories: (1) acidic metals solutions; (2) alkaline metals etchant solutions; (3) cyanides; and (4) chelated metals solutions. The case study for this facility focuses on the use of a sludge minimizing treatment technology for the metals and cyanides wastes.

Initially, the facility was designed to operate using lime and ferrous sulfate precipitation of metals as the primary means of waste treatment. When the high cost of land disposal of the lime sludges was considered, alternate means of treating and disposing of the waste were evaluated.

The unit processes selected to detoxify the wastes and recover metals at Plant A currently include sodium hypochlorite oxidation of cyanides (alkaline chlorination), sodium hydroxide precipitation, pH adjustment, sodium borohydride reduction (with sodium metabisulfite stabilization), sedimentation, plate and frame filter press (for sludge dewatering), rapid sand filtration, and ion exchange columns for effluent polishing.

Results--The primary purpose of the Facility A case study was to evaluate sodium borohydride as a viable waste treatment alternative for reducing RCRA Hazardous Waste Code F006 spent electroplating baths. The evaluation criteria were the ability of sodium borohydride (SBH) to effectively meet local compliance standards and produce a high density, low-volume sludge. The test program evaluation relies mainly on the trace metals results to evaluate system performance.

The SBH reactor was sampled for trace metals on the influent, effluent, and sludge streams. Both filtered and unfiltered samples were collected and analyzed for eight selected metals. The unfiltered sample showed little or no reduction as expected. However, the filtered sample showed individual metals reduction efficiencies which ranged from 16.1 to 99.8 percent. The observed range in efficiency data was attributed to variations in concentration and chemical potential (quantity of free energy required for an ionic species to obtain equilibrium) of each of the metallic ions contained in the solution. Overall, SBH was able to reduce 6.91 kg of the initial influent metals loading of 7.25 kg. These results represent a greater than 95 percent reduction in total metals for a complex waste stream. The remainder of the metals influent loading (0.337 kg) consisted of over 70 percent calcium.

An additional objective of this program was to evaluate the ability of Facility A to consistently meet local pretreatment requirements. The resultant data for two separate batch runs showed discharges in excess of effluent limits, apparently due to incomplete polishing caused by cation exchange column breakthrough. Since the test program was completed, Facility A has instituted the use of a quality control holding tank and further waste processing optimization to remedy these problems. Follow-up discussions with the local sewer authority revealed that Facility A's effluent quality has improved considerably and is now consistently meeting compliance guidelines.

In addition to assessing wastewater effluent characteristics, the testing program was designed to evaluate uncontrolled process air emissions. The results were obtained by Draeger tube analysis of grab and integrated samples of exhaust gases taken from the process reactor exhaust ducts. The emission results showed a frequent presence of hydrochloric acid and hydrogen gas accompanied by occasional presence of ammonia and sulfur dioxide. One of the hydrogen emissions grab sample results (6.0 percent) is significant since this value is greater than the lower flammable limit for hydrogen (4.0 percent). Grab sample concentrations for ammonia and sulfur dioxide also exceeded adopted short-term exposure limits (STEL) for these substances.

Analysis of the nickel/cyanide and SBH sludges shows total metals contents (dry weight) of approximately 35 and 6 percent, respectively. Neither sludge result supported Facility A's claim of 60 to 70 percent metals content (dry basis). While the SBH sludge result was significantly below performance expectations (70 percent metals), the exact cause of these results was not discernable. Possible explanations include: (1) a possible process upset; (2) sampling error; and (3) analytical error. It seems most probable that a process upset was responsible for these results, since blinding of the sludge press occurred on the SBH press. Based on other SBH reduction case study results conducted under this program, it is reasonable to assume that these results are not representative, since typical sludge metals contents should be greater than 70 percent.

EP Toxicity analyses were also conducted for both the nickel/cyanide and SBH reactor sludges. The results of the tests clearly show that for Facility A influent metals concentrations, the SBH sludge produced is fairly stable in that its leachate characteristics are below EP Toxicity limits for all metals. However, note that the waste is still classified as F006 hazardous waste.

An additional objective of the Facility A case study was to evaluate the ability of sodium borohydride to economically reduce F006 waste streams. At the time of testing, Facility A reduction chemistry was very inefficient at $19.80/lb of copper reduced. However, through process optimization, chemical costs have reportedly decreased over 63 percent, bringing process economics within acceptable limits. The case study follow-up for Facility A has indicated that the cost of copper reduction has been lowered to $7.27/lb of copper.

Facility B Case Study--

    Description--Facility B is a captive printed circuit board manufacturing facility employing 77 people in Santa Ana, California. Gross sales are approximately $7 million annually on production of 500,000 ft^2 of board. Production at Facility B uses a special hybrid process, employing elements of both additive and semi-additive printed circuit production techniques. Process wastes of interest to this study include rinsewaters from the electroplating and etchant baths. The principle components of the acid copper electroplating baths are copper sulfate and sulfuric acid. Facility B uses a slower acting etchant (sodium chloride, sodium chlorate, and muriatic acid) which etches copper from the board, and yields cupric chloride in the waste stream.

    Facility B uses a rather unique end-of-pipe treatment system employing sodium borohydride treatment and ultrafiltration (Memtek) technology for solids separation. In this process, incoming plating and etching wastes are adjusted to pH 7-11 by addition of sodium hydroxide or sulfuric acid. Sodium borohydride is added to obtain an oxidation reduction potential (ORP) of approximately -250 mv or less. The reacted waste then feeds from the concentration tank to a Memtek ultrafiltration unit from which the permeate is discharged to municipal treatment, and the concentrate is returned to the concentration tank. A small plate and frame sludge filter press dewaters the sludge which is drawn from the bottom of the concentration tank.

    Points of interest in evaluating the Facility B waste treatment system for this case study were: (1) compliance of the ultrafiltration permeate (wastewater discharge) with local and Federal discharge standards; (2) the volume and EP toxicity of the sludge filter cake; and (3) economic evaluation against comparable technology (lime and ferrous sulfate treatment).

    Results--The objective of the sampling program was to evaluate the effectiveness of the sodium borohydride technology in use by Facility B. The effectiveness was measured in terms of metal reduction efficiency and minimization of hazardous waste streams. Data derived from the metals concentrations in the influent and effluent streams were used to determine the effectiveness of the SBH reduction system in both meeting effluent guidelines and minimizing releases to the environment.

    Analysis of the influent and effluent streams metals characteristics, showed that copper was reduced most efficiently (99.82 percent), while nickel reduction was the least efficient at (45.5 percent). Differences in removal efficiencies were attributed to variations in concentration (higher removals for higher concentrations), but the chemical potential may also have been a factor. Approximately 144.7 lbs of combined metals were reduced to elemental form by the SBH reaction system, representing a combined reaction efficiency of 99.8 percent. Despite deviations from design operating conditions, the SBH/ultrafiltration system performed very well. EP Toxicity leachate test results for Facility B filter press sludge clearly show that the sodium borohydridge sludge produced is fairly stable. Leachate characteristics are below EP Toxicity limits for all metals. However, note that the waste is still classified as F006 hazardous waste.

In addition, an economic comparison of the use of sodium borohydride versus lime-ferrous sulfate chemistries was conducted. The results demonstrate that in this application, sodium borohydride would be superior to lime-ferrous sulfate for the following reasons: (1) sludge disposal costs and volumes would be reduced by 93.5 percent; (2) overall operating expenses would be 48 percent lower; and (3) sludge generated by the SBH reduction process was 78 percent copper and suitable for reclamation (due to the high copper content).

The use of the sodium borohydride and ultrafiltration treatment at Facility B is favored by the use of the chloride etch process in lieu of the more commonly preferred ammonium peroxide etch. The ammonium-based etchants create borohydride sludge stability problems which require tighter treatment process control and the use of stabilizers such as sodium metabisulfite. Additional factors which favor the economics of sodium borohydride treatment at Facility B include: (1) the use of cupric chloride etchant; (2) high copper concentrations and low organic loadings seen at this facility; and (3) low effluent limitations required by the sanitation district.

Based on the above results, it appears that sodium borohydride reduction is an effective technology which can be utilized to reduce complex metal electroplating sludges and render them reclaimable, and possibly less hazardous. Note that the economics of SBH technology is highly dependent on site-specific factors and warrants a detailed study prior to implementation.

Facility E Case Study--

Description--Facility E began operations in January 1982 as a manufacturer of customized, fine-line multilayer printed circuit boards. Facility E initiated an ambitious waste minimization program in mid-1984. Since that time, production has roughly doubled, but liquid discharge to the wastewater treatment plant has remained constant and wastewater sludge generation has dropped roughly 30 percent. Waste minimization efforts continue to center around in-process modifications to use nonhazardous or reclaimable solutions, to reduce water consumption and bath dump frequency, and to optimize wastewater treatment operations.

At Facility E, boards are pattern plated with eight acid copper and one aqueous tin/lead plating baths in a 48-tank plating line. The line begins with a nitric acid (HNO_3) rack strip tank. After the racks are stripped, boards are loaded and then undergo rinsing, cleaning with phosphate solutions (H_3PO_4, Electroclean PC2000), and more rinsing before being plated. Acid copper baths contain $CuSO_4$, organic brighteners, and chlorides with copper concentrations of 24 oz/gal. The general processing procedure is to activate the board surface (HCl), plate, clean/rinse and replate.

In plating operations, addition agent and photoresist breakdown products will incrementally accumulate and contaminate an electrolytic (charge carrying) plating bath. In the absence of a bath regeneration system, the manufacturer would typically be forced to either discharge the spent plating bath to the wastewater treatment plant or send it offsite for disposal. In either case, large quantities of metals containing sludge (RCRA Waste Code F006) would be generated and subsequently land disposed. At Facility E, these spent plating baths are regenerated through activated carbon filtration (used to remove built-up organic bath contaminants) and then returned to the process. Copper and solder plating baths are treated with activated carbon once every three months and every month, respectively. The frequency of cleaning is determined by organic contaminant build-up. Electroplating baths never have to be dumped with this arrangement under normal processing conditions.

Activated carbon treatment is performed in a batch mode for acid copper, solder and nickel microplating baths in three separate systems. The bath reclamation system consists of a holding tank, mixing tank, and MEFIAG paper-assisted filter. For acid copper treatment, 2,400 gallons of contaminated solution is pumped into a 3,000 gallon mixing tank. Hydrogen peroxide is added and the temperature of the bath is maintained at 120 to 130°F for 1 hour. Powdered activated carbon (80 lbs) is added and the contents are mixed for 3 to 4 hours to oxidize volatile organic species. The solution is recirculated through a paper-lined MEFIAG filter several times to remove the activated carbon. The filter solids and paper are removed as needed when a predetermined pressure drop across the filter is reached. When the bulk of the activated carbon has been removed (generally after three passes of the solution through the filter), the filter is precoated with 5 gallons of diatomaceous earth. The solution is again recirculated through the filter until a particulate test indicates sufficient solids removal (no residue detected on visual examination of laboratory filter paper). Total spent solids from plating bath purification is 1-1/2 drums every 3 months which is landfilled.

Results--The purpose of this case study was to evaluate the extension of electroplating bath lifetimes (and subsequent waste reduction) by activated carbon removal of organic brightner breakdown products. The acid copper baths were selected for study since recovery of this solution results in the most significant amount of waste minimization.

Sampling and analysis was conducted on three process streams associated with activated carbon bath reclamation. Based on resultant analytical data the following conclusions were drawn:

●   Forty-seven percent of the organic by-products and brightners were removed from the contaminated solution;

- Low molecular weight organics such as carboxylic acid derivatives are not preferentially adsorbed;

- Reduced sulfur (a brightening and leveling agent) is oxidized and volatilized during treatment; and

- Inorganic contaminants such as tin and lead are also removed (37.5 percent and 24.5 percent, respectively) as a beneficial by-product of the treatment process.

In recovering spent electrolytic plating baths, Facility E was able to save over $50,000 in hazardous waste disposal and raw material purchase costs. These savings represent a payback period of only 3 months for purchasing the activated carbon recovery system. This relatively short payback period, combined with the volume of plating solution regenerated, make activated carbon treatment a cost-effective and environmentally safe technology for reducing the quantity of hazardous waste that would otherwise be land disposed.

Facility F Case Study--

Description--Facility F is an independent manufacturer of printed circuit boards. The normal production volume of the facility is 40,000 ft^2/month. The major waste streams of interest to this case study are rinsewaters that follow electroplating and etching processes. Prior to implementation of the electrolytic recovery technology being studied, these rinsewaters contained copper and lead at concentrations of up to 3,000 mg/L. Because of this, the concentration of these metals in the final effluent exceeded pretreatment standards (4.5 mg/L for copper and 2.2 mg/L for lead) for discharge to the city sewer system. To decrease the concentration of metals in the effluent, the facility converted several rinse tanks into static dragout tanks in order to recover metals from rinse baths following copper electroplating, tin/lead electroplating, electroless copper plating, and a copper microetch process. The quantity of metal recovered from the electroless copper rinse and the copper microetch was small. Thus, the reactors were removed from these baths and installed at the copper and tin/lead rinse baths where there was more potential for metal recovery.

The electrolytic reactors used at this facility are Agmet Equipment Corp., Model 5200 reactors. They consist of a wastewater sump, a pump, and the anode and cathode, contained within a rectangular box with dimensions of approximately 22 in. x 10 in. x 22 in. The anode was cylindrical and was encircled by a stainless steel cathode with a diameter of 8 in. and a height of 6 in. The anode material used for copper plating solutions is titanium. For tin/lead plating solutions, however, a columbium anode was required because the fluoroboric acid in the tin/lead plating solution was extremely corrosive to titanium. The columbium anode increases the cost of these electrolytic units to $4,500, as opposed to $3,500 for the titanium anode units.

Results--At the time of testing, four electrolytic reactors were being used for recovery of copper, and three were being used for recovery of tin/lead. To evaluate the performance of these units, samples of the plating bath, dragout, and rinse bath were analyzed. Conclusions that were drawn based on the resultant data include:

- Recovery of copper from the acid copper solution is very effective-- rates of recovery were 4 to 5 grams/hour/unit, representing a current efficiency of nearly 90 percent.

- Recovery of tin and lead was not effective at the time of testing-- concentrations of these two metals in the dragout were not significantly less than in the plating bath. However, evaluation of the data was difficult because the analytical results for some of these samples were inconclusive due to matrix interference.

- Use of in-line electrolytic recovery was not able to reduce metal concentrations sufficiently as to enable this facility to meet pretreatment standards.

- Electrolytic recovery would significantly reduce the amount of sludge generated if a lime precipitation system were utilized to remove metals from the final plant effluent. For this facility, a reduction of 32 tons/year would be realized.

- At a sludge disposal cost of $200/ton, the annual cost of electrolytic recovery would exceed the savings. However, if sludge disposal costs increased to $300/ton, the savings (at least for copper recovery) would exceed the processing costs.

Electrolytic recovery methods remove metals from an aqueous solution in a metallic form which allows for the use of the recovered material as scrap metal. Conversely, hydroxide precipitation removes the metal from solution and generates a sludge with a low metal concentration. In most cases, the only method of handling this sludge is landfilling at a high cost. Therefore, electrolytic recovery is useful in minimizing the quantities of metal-bearing sludge that must be landfilled. The cost effectiveness of this type of technology will increase as sludge disposal costs increase in the future.

Resist Developing Solvent Recovery Case Studies

Two case studies evaluated under this program focused upon the minimization of developer solvent wastes and sludges which might require either land disposal or incineration. In general, the recovery of resist stripping and developer solvents is not unique within the PC board manufacturing industry. However, the recovery systems evaluated at the two facilities discussed below represent state-of-the-art technology applications. In the case of Facility C, the technology involves the separation of a two-solvent system with subsequent recovery and reuse of each solvent. In the case of Facility D, the technology evaluated further recovers the solvent bottoms product of the initial recovery unit.

Facility C Case Study--

Description--Facility C manufactures computing equipment including logic, memory and semiconductor devices, multilayer ceramics, circuit packaging, intermediate processors and printers. One of the major hazardous waste streams that is generated is spent halogenated organic solvents (RCRA Code F002). Methylene chloride is used in resist stripping of electronic panels. Methyl chloroform (1,1,1-trichloroethane) is used in resist developing of electric panels and substrate chips. Freon is used in surface cleaning and developing of substrate chips. Perchloroethylene used in surface cleaning of electronic panels.

The spent solvents from photoresist stripping and developing are contaminated with photoresist solids at up to 1 percent, and the solvents used for surface cleaning are contaminated by dust, dirt or grease. Waste solvents are recovered at Plant C by distillation or evaporation and returned to the process in which they were used. Several types of equipment are used including box distillation units to recover methylene chloride and perchloroethylene, flash evaporators to recover methyl chloroform, and a distillation column to recover Freon.

There are two identical flash evaporators at the facility, each with a capacity to recover 600 gallons of methyl chloroform (MCF) per hour. The flash chamber operates at a vacuum of 20 in. Hg, allowing the MCF to vaporize at 100 to 110°F. The units are operated one to two shifts/day depending on the quantity of waste solvent being generated.

A packed distillation column is used to recover pure Freon from a waste solvent stream containing approximately 90 percent freon and 10 percent methyl chloroform. Waste is continuously fed to a reboiler where it is vaporized and rises up the packed column. Vaporized freon passes through the column, is condensed and recovered at a rate of 33 gal/hour. MCF condenses on the packing and falls back into the reboiler. The distillation bottoms are removed when the concentration of methyl chloroform reaches 80 percent (approximately 1 to 2 weeks).

There are also two identical box stills at the facility, each with a capacity to recover 475 gph of methylene chloride. These are very simple units consisting of an 800 gallon still pot with hot water heating coils. The contaminated methylene chloride is heated to between 103°F and 108°F, and clean solvent is condensed overhead.

Results--Sampling and analysis was conducted on process streams associated with two of the solvent recovery processes. One of these processes was the flash evaporator used for recovery of methyl chloroform (1,1,1-trichloroethane), and the other was the distillation column used to recover Freon TF from a Freon/methyl chloroform mixture. The conclusions drawn from the sampling and testing program were:

● At least 95 percent of the solids are removed from the solvent waste influent;

- The recovered product is at least as clean as the virgin material; and

- The still bottoms from recovery of contaminated solvent still contain a high fraction (90 percent) of solvent.

The recovery of spent solvents at the facility is motivated primarily by economic benefit. In recovering spent solvent, the company saves over $10 million annually, compared to offsite recovery. The savings per pound of methyl chloroform, methylene chloride, and Freon recovered is $0.18, $0.18, and $0.61, respectively.

The high cost savings are primarily due to the fact that the solvents recovered are reused onsite, thus reducing the quantity (by greater than 95 percent) of new or virgin solvent that must be purchased. Offsite recovery could be conducted, but at much higher cost. Since the rate of generation of spent solvent is so high, the initial expense of purchasing recovery equipment is quickly returned.

To landfill or dispose of such a large quantity of spent solvent by any other method would be economically unacceptable. Incentives other than economic reasons for onsite recovery include:

- Reduction in the risk of a spill of solvent in transporting the waste to a TSDF; and

- Reduced liability related to an accident at the TSDF resulting in the release of spent solvent.

Facility C is trying to further reduce the quantity of waste solvent that must be sent offsite for recovery. They intend to do this by recovering the still bottoms generated by distillation of Freon/methyl chloroform waste. In addition, they eventually plan to phase out the use of methyl chloroform and methylene chloride and replace these materials with aqueous-based photoresist developers and strippers.

Facility D Case Study--

Description--Facility D manufactures mobile communications equipment components in their Florence, S.C. facility. The operation consists of a small metal-forming shop, prepaint and painting lines, electroplating, printed circuit board manufacture, and a 30,000 gpd onsite wastewater treatment plant.

Printed circuit boards are produced using the subtractive technique and solvent-based photoresists. Methylene chloride resist stripper and 1,1,1-trichloroethane (TCE) developer are continuously recycled in closed-loop stills. The TCE developer wastes (Waste Code F002) are recovered in a DuPont Riston SRS-120 solvent recovery still (referred to as the primary still) and returned to the developer line. Until recently, all still bottoms from the primary still were drummed and shipped offsite for reclamation at a solvent recycling facility. Facility D purchased a Recyclene Industries RX-35 solvent recovery system (referred to as the secondary still) in October 1985, to further remove TCE from still bottoms onsite.

The Recyclene Industries RX-35 solvent recovery system is a batch distillation system with a 30 gallon capacity, silicone oil immersion heated stainless steel boiler, a non-contact, water-cooled condenser, and a 10 gallon temporary storage tank. The boiler is equipped with a vinyl liner inside a Teflon bag. The Teflon bag provides temperature resistance and the vinyl bag collects solid residue, eliminating boiler clean-out and minimizing sludge generation after distillation. Two thermostats control the temperature of the boiler and the vapor, automatically shutting down the boiler when all the solvent has evaporated. The maximum operating temperature of the still is 370°F, so recovery of solvents with higher boiling points would not be practical. Recovery of a 20 to 25 gallon batch of still bottoms requires approximately 90 minutes at Facility D, and four to six batches are completed each day.

Results--Evaluation of the system consisted of the analysis of the contaminated feed, overhead product, and distillation bottoms. Based on a mass balance and analytical data, the following conclusions were made:

- Purity of recovered solvent was 99.99 percent;

- Total solvent recovery was 99.78 percent;

- Still bottoms contained 7.5 weight percent 1,1,1-trichloroethane; and

- Reduction in waste generation was 97.5 percent..

An additional objective of the study was to evaluate the economics of the batch solvent recovery unit. Annual cost savings ($43,000) and waste reduction (110,602 gal) were calculated for Plant D, based on the first year of RX-35 operation. In addition, the investment payback period for the RX-35 was calculated considering credit for reclaimed solvent and reductions in waste transportation and disposal costs. The estimated payback period was 7.3 months, given the current level of solvent reclamation. Thus, the low capital cost of the unit and the relatively high costs of virgin solvent ($4.50/gal) favor the second-stage recovery of TCE developer still bottoms.

There are several potential drawbacks to the implementation of RX-35 batch still that should be discussed. The first is that since the bottoms product contains 7.5 weight percent 1,1,1-TCE, it remains classified as RCRA Waste Code F002 (halogenated organic solvents) and is among those solvent wastes being considered under the land disposal ban. Thus, while this technology significantly reduces the volume and toxicity of the solvent still bottoms, it continues to generate a hazardous waste. A second potential concern is the accumulation of contaminants and/or breakdown products. For example, 6.7 to 11.0 percent concentrations of carbon tetrachloride were found in process feed and exit streams, indicating a build-up of this contaminant. Another significant contaminant found was 2-Butanone, which represents 3.6 percent of the solvent waste feed stream. It could not be determined whether a build-up of 2-Butanone was occurring or if it is harmful to the system. However, its presence and effect on the solvent properties of 1,1,1-TCE should be considered.

A final consideration in the implementation of any solvent recovery still is the issue of safety. The unit at Plant D was housed in a separate structure and provided with adequate ventilation to minimize the risk of exposure or explosion. The RX-35, according to the manufacturer, is safe for flammable materials, and is rated for NFPA Class 1, Division I, Group D environment (Recyclene, 1985). These safety considerations should help to minimize the risk of chronic exposure or danger from explosion to personnel. Nevertheless, explosion risks from solvent recovery operations should be carefully evaluated in planning the layout and installation of the unit.

2. Conclusions

ELECTRONIC INDUSTRY WASTE MANAGEMENT

In the manufacture of printed circuit boards and semiconductors, major waste streams of concern are spent organic solvents (RCRA codes F001-F005) and metals-containing wastes and wastewater sludges (RCRA code F006).

Organic Solvent Wastes

Organic solvents are used for wafer/board cleaning and for the developing and stripping of photoresist materials used in the image transfer and/or circuit fabrication processes. The electronics component industry ranks high relative to other industries in the generation of solvent waste. Semiconductor manufacturers are ranked 12th and electronics component manufacturers not elsewhere classified (which includes the manufacture of printed circuit boards) are ranked 19th. As companies continue switching to photoresist materials with an aqueous or semiaqueous base as opposed to an organic solvent base, quantities of organic hazardous waste generated by this industry should decrease. However, many companies will continue to employ the solvent-based process due to the high capital costs associated with conversion. For these companies, onsite waste reduction will become an important means to reduce waste treatment costs and future liabilities. Thus, onsite solvent still bottoms recovery will see increasing prevalence as land disposal costs and offsite processing costs continue to rise.

Since most spent organic solvents are still quite valuable, recovery has been a common method of management. Solvents used in the electronics industry require a high purity which is difficult to achieve by standard solvent distillation practices. Consequently, it is easier to send these wastes offsite where the majority of the contaminants can be removed, and the recovered solvent can be used in an application requiring lower solvent purity.

Recovery of solvents by distillation results in the generation of a bottoms product containing contaminants and up to 95 percent of the organic solvent. Secondary recovery of the solvents is often possible through the use of supplementary technologies such as steam distillation or thin film evaporation. These methods significantly reduce waste product stream volume and represent feasible and readily implemented methods of hazardous waste management.

Metals-Containing Wastes

Metals are essential to all electronic components due to their conductive and resistive properties. The most common forms of application are electroless and electrolytic plating, in which an adherent metallic coating is deposited on an electrode (the part being plated) to produce a surface with properties or dimensions different from those of the basic metal. These metals are introduced into the waste stream through either the disposal of concentrated plating baths or running rinses directly following the electroplating process. A second major source of metallic contaminants is the chemical etch step utilized as part of the electroplating preclean operations or in the removal of excess surface metal. Etching rinses will contain relatively high concentrations of metals along with dilute levels of etching solution. Conventional waste treatment for metals containing waste includes chemical precipitation, clarification, and dewatering, which results in the landfilling of hazardous sludges (RCRA code F006).

As effluent discharge limits for the electronics industry have become increasingly strict, the industry has been forced to treat their wastewaters to remove dissolved metals. However, conventional treatment methods such as lime precipitation results in the generation of large quantities of metal containing sludges. Since disposal of these sludges in landfills may soon be banned under the amendments to RCRA, other nonsludge generating methods of management will see increasing utilization.

Offsite use, reuse, recovery or recycle (URRR) consists primarily of sending spent plating and etching solutions back to the manufacturer of these solutions to be regenerated. Onsite recovery processes, however, such as the electrolytic recovery of metals from rinsewaters, has yet to achieve widespread use. Methods for onsite reduction of the quantity of hazardous metals-containing sludge include sodium borohydride reduction, ion exchange, electrolytic recovery, evaporation, reverse osmosis, and electrodialysis. These techniques for recovering metals from wastewaters have become more common since 1981 and new methods are constantly being developed.

CASE STUDY FINDINGS

The findings of the six waste minimization case studies tested under this program are presented in Table 2, which include data collected by the facilities and verified by sampling and laboratory results. These results indicate that a variety of technologies exist to minimize metals-containing and solvent wastes produced by the printed circuit board and semiconductor industries. The technologies discussed range from simple changes in treatment system reagents with nominal capital costs to large onsite solvent reclamation facilities with significantly higher capital costs.

Four of the case studies investigated under this program focused on technologies to reduce metal-plating rinsewater sludges. Two of the case studies, evaluating the use of sodium borohydride reduction as a substitute for lime/ferrous sulfate precipitation, found that the technology was a viable substitute in one case and appeared to be marginally acceptable in another.

TABLE 2. SUMMARY OF FINDINGS OF WASTE REDUCTION CASE STUDIES

Facility name	Technology	Waste reduction	Annual Waste reduction achieved	Capital costs ($)	Projected annual cost savings ($)
Facility A	Sodium borohydride reduction	Metals sludge	--a	Nominal	--b
Facility B	Sodium borohydride reduction	Metals sludge	962 tons	Nominal	115,870
Facility C	Solvent batch distillation	Methylene chloride Methl chloroform Freon	6,152,000 gal	709,400	16,000,000
Facility D	2-Stage solvent distillation	1,1,1,-Trichloroethane Resist developer still bottoms	10,625 gal	26,150	43,105
Facility E	Carbon adsorption plating both reclamation	Plating bath wastes (metals sludge)	10,600 gal	9,200	57,267
Facility F	Agmet electrolytic recovery unit	Metals sludge	32 tons	30,350	(10,685)c

aNot quantifiable, but a significant waste reduction was realized.
bNot demonstrated during testing.
c( ) indicates negative value.

The case study on carbon adsorption removal of harmful organic contaminants from plating bath wastes found that this technology significantly reduced both disposal costs and waste volume. The case study of electrolytic recovery indicated that this technology is highly waste stream specific. An acid copper electroplating rinse is an ideal waste stream for electrolytic recovery. However, other metal-bearing rinses, such as those from solder (tin/lead) plating, or etching are not appropriate for use of electrolytic recovery. Electrolytic recovery units are, however, generally inexpensive to purchase and can be used in many cases to supplement an end-of-pipe treatment process.

Two of the case studies presented in this program involved the recovery of spent halogenated solvents using batch distillation units. Both of these case studies indicate that onsite solvent recovery is successful from a technical and an economic standpoint. In both cases, over 95 percent of the waste solvent was recovered and reused onsite. Solvent recovery appears to be a technology that can be applied to a number of printed circuit board manufacturing facilities.

The results of this project indicate that waste reduction can be achieved through the use of an appropriate technology, and it can be achieved with significant reductions in cost. The case studies also indicate that the success of waste reduction is in many cases waste stream specific. The technologies will not necessarily be successful in all cases. A slight variation between one waste stream and another may make waste reduction either technically or economically impractical. Therefore, successful waste reduction is dependent on a thorough knowledge of waste quantities and characteristics.

3. Recommendations

As the case studies presented in this document indicate, cost-effective application of waste reduction technologies is dependent on site specific factors such as waste volume, waste characteristics, and availability of existing onsite facilities and technical expertise. The latter is particularly lacking in small businesses which often do not possess specially trained personnel that are able to devote the time required to investigate waste treatment options. Due to this factor and economies of scale, these businesses currently land dispose a disproportionately high percentage of their wastes whereas large quantity generators are more apt to employ waste minimization and recycling practices. Thus, the land disposal restrictions and consequently dissemination of waste reduction information, will have a more significant impact on smaller waste generating firms. This is particularly true now that the small quantity generator exclusion limit has been lowered.

Industries will also be impacted to varying extents based on the type of wastes they generate and the effective dates for promulgation of the land disposal restrictions for these wastes. Solvent wastes, with total organic content of one percent or more, are the first waste types to be banned from land disposal, effective November 8, 1986. Industries which currently land dispose large quantities of these wastes include a wide range of small volume generators including metal finishers, electronic component and equipment manufacturers, and dry cleaners. In addition, these industries consume relatively large quantities of halogenated solvents. Since these wastes tend to be generated in smaller volumes, are more restricted in terms of available disposal options, and are more expensive to purchase relative to their non-halogenated counterparts, they are particularly well suited to the application of waste minimization and recycling technologies. Thus, future EPA information dissemination should focus on substitutes and recovery and treatment alternatives for halogenated solvents. In particular, performance data are lacking for high solids processing units, like the Recyclene distillation unit used at Facility D, and disposal options for the resulting residuals.

Residual disposal costs will represent an increasingly important factor in overall system cost-effectiveness as the scale of the operation increases. This is particularly true for chlorinated or metals containing residuals with moderate organic contents. These are expensive to incinerate and are not amenable to conventional stabilization/encapsulation techniques. Additional guidance on optimal treatment process selection and research on alternative residual disposal methods is required to assist generators of these wastes. Although large generators are likely to be impacted less severely than small generators as a result of the land disposal ban, it must be recognized that they are responsible for the majority of waste generation and disposal. Thus, research which is oriented towards the management of large quantity generator wastes will result in the greatest overall reduction in waste disposal costs and its associated environmental hazard.

Candidate technologies which appear promising but for which performance data are currently limited include chemical fixation, encapsulation, use as a fuel substitute in aggregate kilns and blast furnaces, and dechlorination techniques for halogenated solvents. Non-halogenated organics are more ammenable to conventional thermal destruction techniques. Similarly, other waste types (e.g., corrosives, metal bearing sludges) are also amenable to conventional disposal techniques (e.g., neutralization, stabilization/ encapsulation) and thus will not be subject to as high an increase in disposal costs as can be expected for halogenated organics. However, the large volume of these wastes justifies further research and information dissemination to assist industry in complying with the land disposal ban in the most cost-effective manner. In particular, additional performance data are required for membrane and other metal recovery technologies that can withstand corrosive environments such as that found in many pickling, etching and plating baths.

In summary, EPA activities to date in this and other programs have focussed on the identification of and dissemination of information on waste reduction and treatment technologies. This effort has served to inform industry of current cost-effective practices and to identify wastes for which currently available data are lacking. Future efforts should target specific wastes which create the most significant disposal problems in terms of overall cost to industry and severity of impact on specific industries. In addition, research should focus on those technologies which are most likely to result in cost-effective compliance with the land disposal ban regulations.

4. The Electronics Products Industry

The electronic components manufacturing industry (SIC 367), includes eight specific product areas identified by four-digit SICs. These product areas include capacitors, transformers, semiconductors, and printed circuit boards. As determined by the case study selection criteria, the semiconductor and printed circuit board industry were assessed as the product areas of greatest interest. Total worldwide production of printed circuit boards was approximately $4.5 billion in 1984, but has declined by 40 percent in 1985 (Card, September 1, 1985). Worldwide production of semiconductors also experienced a setback in 1985 as evidenced by the 1984 production of $33 billion down to $29 billion in 1985 (Electronic Business, March 1, 1986). However, total U.S. production of semiconductors is forecasted to experience growth from $8.3 billion in 1985 to $15.9 billion in 1988 (Patterson, October 14, 1986), while world production of printed circuit boards will reach $9 billion in 1989 (Card, September 1, 1985).

The industry consists of both small, independent job shops with limited product lines to large automated facilities with integrated operations generating large quantities of hazardous waste. In 1980, there were reported to be 545 companies in the U.S. involved in the manufacture of semiconductors, and 345 involved in the manufacture of printed circuit boards. Only 12 percent of the companies surveyed employ over 2,400 persons, while 80 percent employ 100 or less (EPA, 1983). Due to the high degree of design diversity within product areas and the large disparity between generator volume, wastes are categorized by the primary constituent of the waste, not by raw material usage or manufacturing process.

WASTE GENERATION

In the manufacture of printed circuit boards (Figure 1) and semiconductors (Figure 2), major waste streams of concern are spent organic solvents (RCRA codes F001-F005) or metals containing wastes (characterized in Tables 3 and 4). Organic solvents are used for wafer/board cleaning and for the developing and stripping of photoresist materials used in the image transfer and/or circuit fabrication processes. Photoresists are light sensitive, organic, thermoplastic polymers available as either liquids or dry solids. Negative image photoresists polymerize upon exposure to light, after which unexposed areas are dissolved by developer solvent. Developers and strippers for this type of resist are generally organic solvents such as

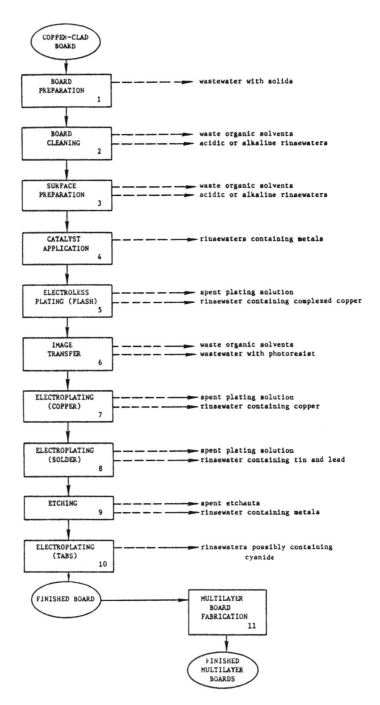

Figure 1.   Subtractive printed circuit board production flowsheet.

Source: EPA-600/2-83-033.

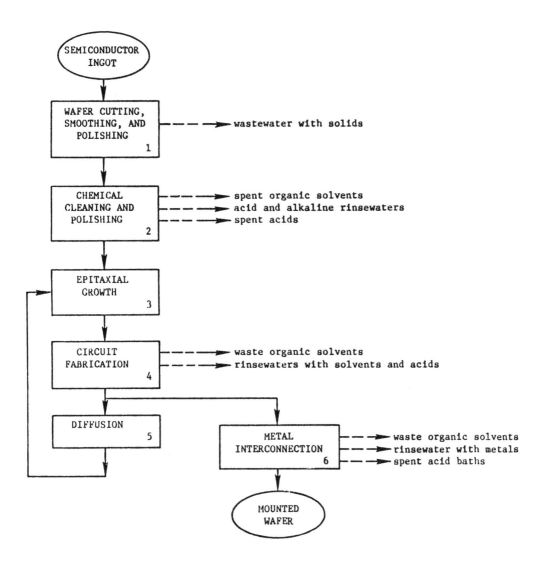

Figure 2.   Integrated circuit production flowsheet.

Source: EPA-600/2-83-033.

TABLE 3.   CHARACTERISTICS OF RAW WASTE STREAMS FROM SEMICONDUCTOR
DEVICE MANUFACTURING (EPA-600/2-83-033)

Parameter	Concentration range, mg/liter	Mean concentration, mg/liter	Industry wide pollutant discharge, kg/day[a]
Antimony	<0.001-0.187	0.021	13.2
Arsenic	<0.003-0.067	0.018	13.2
Beryllium	<0.001-<0.015	0.002	1.9
Cadmium	<0.001-0.008	0.003	1.9
Chromium	<0.001-1.150	0.129	99.9
Copper	<0.005-2.588	0.570	540.7
Cyanide	<0.005-0.01	0.005	3.8
Lead	<0.04-1.459	0.145	61.5
Mercury	<0.001-0.051	0.004	5.7
Nickel	0.005-4.964	0.502	655.6
Selenium	<0.002-0.045	0.021	6.9
Silver	<0.001-0.013	0.005	3.8
Thallium	<0.001-0.012	0.015	11.3
Zinc	0.001-0.289	0.093	46.5
Phenols	<0.002-6.1	0.630	812.6
Oil and grease	ND-20.8	5.058	2,778.3
Total suspended solids	ND-203	31.61	30,470.6
Total organic carbon	ND-80	55.676	17,094.2
Biochemical oxygen demand	9-202	52.768	38,848.1
Fluoride	ND-330	62.0	35,909.0
1,2,4-trichlorobenzene	<0.01-27.1	4.643	257.5
1,1,1-trichloroethane	<0.01-7.7	1.395	928.2
Chloroform	<0.01-0.05	0.015	15.7
1,2-dichlorobenzene	<0.01-186.0	15.972	499.3
1,3-dichlorobenzene	<0.01-14.8	1.450	174.0
1,4-dichlorobenzene	<0.01-14.8	1.341	156.4
1,1-dichloroethylene	<0.01-0.071	0.029	9.4
2,4-dichlorophenol	<0.01-0.017	0.012	9.4
Ethylbenzene	<0.01-0.107	0.021	6.3
Methylene chloride	<0.01-2.4	0.244	276.1
Naphthalene	<0.01-1.504	0.214	19.5
2-nitrophenol	<0.01-0.039	0.024	27.6
4-nitrophenol	<0.01-0.18	0.061	15.1
Phenol	0.014-3.5	0.519	203.5
Di-n-octyl phthalate	<0.01-0.01	0.01	6.3
Tetrachloroethylene	<0.01-0.80	0.122	363.0
Toluene	<0.01-0.14	0.018	33.9
Trichloroethylene	0.007-3.5	0.322	177.1

[a] Flowrate weighted.
ND - Not detected.

TABLE 4.   CHARACTERISTICS OF RAW WASTE STREAMS FROM PRINTED CIRCUIT
           BOARD MANUFACTURING

Constituent	Range, mg/liter
Total suspended solids	0.998 - 408.7
Cyanide, total	0.002 - 5.333
Cyanide, amenable to chlorination	0.005 - 4.645
Copper	1.582 - 535.7
Nickel	0.027 - 8.440
Lead	0.044 - 9.701
Chromium, hexavalent	0.004 - 3.543
Fluorides	0.648 - 680.0
Phosphorus	0.075 - 33.80
Silver	0.036 - 0.202
Palladium	0.008 - 0.097
Gold	0.007 - 0.190
EDTA	15.8 - 35.8
Citrate	0.9 - 1342
Tartrate	1.3 - 1108
NTA	47.6 - 810

Source: EPA-600/2-83-033.

1,1,1-trichloroethane, methylene chloride, xylene, and ethyl benzene. Positive-image photoresist materials become soluble upon exposure to light, after which developer solvent is used to remove resist material under the transparent areas of the photomask. Developers and strippers for this type of resist are generally aqueous solutions which are either alkaline in nature or contain organic compounds such as glycol ethers and alcohols.

A recent trend in electronics component manufacturing is the switch-over from negative to positive photoresist materials. This is particularly evident in states such as California where the Air Resources Board guidelines will require a 90 percent decrease in the emission of volatile organic compounds (VOCs) by 1987. Since positive photoresists utilize aqueous solutions, their use can aid in the compliance with the new standards (Electronic Business, March 1, 1986). In the absence of spent organic solvents, the aqueous solution can be released to the sewer, with solids removal being the only required treatment.

The electronics component industry ranks high relative to other industries in the generation of solvent waste (as shown in Table 5). Semiconductor manufacturers are ranked 12th and electronics component manufacturers not elsewhere classified (which includes the manufacture of printed circuit boards) are ranked 19th. These data, however, reflect 1981 practices. If an increasing number of companies continue switching to photoresist materials with an aqueous or semiaqueous as opposed to an organic solvent base, then the quantity of hazardous waste should be decreasing.

Metals are essential to all electronic components due to their conductive and resistive properties toward electricity. Silver, gold, copper, tin, and their alloys are utilized because their high conductivity is essential to the operation of components or because their use in leads and connectors keeps electrical power loss to a minimum. Many metal parts must be protected from corrosion by plating with nickel, silver, gold, or tin/lead. The most common forms of application are electroless and electrolytic plating, in which an adherent metallic coating is deposited on an electrode (the part being plated) to produce a surface with properties or dimensions different from those of the basic metal (WAPORA Inc., 1977). These metals are introduced into the waste stream through either the disposal of concentrated plating baths or running rinses directly following the electroplating process. A second major source of metallic contaminants is the chemical etch step utilized as part of the electroplating preclean operations or in the removal of excess surface metal. Etching rinses will also contain relatively high concentrations of metals along with dilute levels of etching solution. Chemical etch baths typically contain ammonium chloride, ammonium persulfate, or sulfuric acid/hydrogen peroxide as the active ingredient and are applied in either a batch mode or in a conveyorized spray apparatus. Conventional waste treatment includes chemical precipitation, clarification, and dewatering, which results in the landfilling of hazardous sludges (RCRA code F006).

WASTE MANAGEMENT

As effluent discharge limits for the electronics industry have become increasingly strict, the industry has been forced to treat their wastewaters to remove dissolved metals. As mentioned previously, however, conventional

TABLE 5. TOP 20 INDUSTRIES GENERATING SOLVENT WASTES

No. of estab.[a]	SIC Code	SIC description	Weighted[c] number of solvent waste streams	
			Halogenated solvents[b]	Nonhalogenated solvents[b]
2145	2851	Paints & Allied Products	105	1436
1160	2869	Industrial Organic Chemicals	327	654
1529	2821	Plastics Materials	215	536
4287	3471	Plating and Polishing	471	176
541	2833	Medicinal, Botanical Products	137	323
2902	3479	Metal Coating & Allied Serv.	136	279
4656	3662	Communication Equipment	186	225
4151	3714	Motor Vehicle Parts	241	161
393	9711	National Security	166	178
337	3721	Aircraft Equipment	107	230
15490	3079	Plastic Products, Misc.	120	201
2237	3674	Semiconductors	93	194
2563	2899	Chemical Preparations	85	189
6506	7391	Research & Devel. Labs	103	163
560	3411	Metal Can Fabrication	35	154
1040	3711	Motor Vehicle Bodies	57	127
32	2067	Chewing Gum	57	87
861	2879	Agricultural Chemicals	59	85
5392	3679	Electronic Components	96	40
235	3951	Pens & Mechanical Pencils	66	59
57017				

[a]Number of establishments based on Dun's Marketing Services, a company of Dun and Bradstreet Corp., 1983 Standard Industrial Classification Statistics.

[b]Information on generators taken from 1981 data (National Survey of Generators).

[c]For weighting procedure refer to Westat, Inc., 1984.

Source: Engineering Science, 1984.

treatment methods such as lime precipitation result in the generation of
metals containing sludges. Since disposal of these sludges in landfills may
soon be banned under the amendments of RCRA, other nonsludge generating
methods of management will see increasing utilization. Data from the National
Survey of Waste Generators, which reflects 1981 practices, show that SIC 36
(electronics industry) ranks second among all industrial categories in offsite
use, reuse, recovery, or recycle (URRR) of hazardous waste. By contrast,
SIC 36 is not ranked in the top ten for onsite URRR. It is believed that
offsite URRR consists primarily of sending spent plating and etching solutions
back to the manufacturer of these solutions to be regenerated. Data in
Table 6 indicate that this type of practice has been common. Onsite recovery
of metals from rinsewaters has yet to achieve widespread use.

Some of the methods for onsite minimization of the quantity of hazardous
sludge include sodium borohydride reduction, ion exchange, eletrolytic
recovery, evaporation, reverse osmosis, and electrodialysis. These techniques
for recovering metals from wastewaters have probably become more common since
1981, and new methods are constantly being developed.

Most spent organic solvents, with contamination of less than 5 percent,
are valuable, and are recovered for reuse. This is confirmed by the data in
Table 7 which indicate that 40 million gallons of solvent waste listed by
SIC 36 were recycled in 1981. The majority of this 40 million gallons
(70 percent) was recycled offsite. In contrast the chemical manufacturing
industry (SIC 28) employed 87 percent onsite URRR. One conclusion that may be
drawn from this data is that solvents used in the electronics industry require
a high purity which is difficult to achieve by standard solvent distillation
practices. Consequently it is easier to send these wastes offsite where the
majority of the contaminants can be removed, and the recovered solvent can be
used in an application requiring lower solvent purity.

In recovery of solvents by distillation, there is generation of a bottoms
product containing contaminants and up to 95 percent of the organic solvent.
Secondary recovery of the solution is many times possible through the use of
supplementary technologies such as steam distillation or thin film
evaporation. These methods significantly reduce waste product stream volume
(up to 90 percent of the solvent) and combined with positive photoresists and
aqueous based developers and strippers represent feasible and readily
implemented methods of hazardous waste management.

TABLE 6. QUANTITIES OF METAL-CONTAINING WASTE URRR (GALLONS) [Versar, 1985]

Offsite URRR

2-Digit SIC Code	Generator	TSD	Total
33	22,751,971	2,194,224	24,946,195
36	7,953,151	465,598	8,418,749
37	3,090,480	3,164,325	6,254,805
28	2,277,109	851,305	3,128,414
49	2,361,047	1,422	2,362,469
29	1,246,185	1,083,679	2,329,864
34	601,926	44,496	646,422
34	129,872	278,829	408,701
97	0	269,011	269,011
30	0	133,793	133,793

Onsite URRR

2-Digit SIC Code	Generator	TSD	Total
37	463,384,736	202,215	463,586,951
29	29,401,566	52,751,679	82,153,245
33	3,305,980	35,267,641	38,573,621
28	929,989	23,164,213	24,094,202
39	4,357,800	0	4,357,800
50	0	2,591,306	2,591,306
30	0	994,365	994,365
32	0	235,232	235,232
34	155,247	63,429	218,676
31	180,545	0	180,545

TABLE 7.   RECYCLING OF SOLVENT WASTES, LISTED BY SIC[a]

SIC	Waste Volume (gals/yr)[b]		
	Recycled offsite	Recycled onsite	Total recycled
0	1,260,842 (20)	5,151,776 (80)	6,412,618
1	10,691 (100)	0 (0)	10,691
7	593 (100)	0 (0)	593
10	620,232 (100)	0 (0)	620,232
14	0 (0)	4,376,901 (100)	4,376,901
16	9,542 (100)	0 (0)	9,542
17	96,225 (42)	130,661 (58)	226,886
20	58,642 (37)	100,317 (63)	158,959
22	385,388 (52)	351,977 (48)	737,365
23	26,739 (100)	0 (0)	26,739
24	199,085 (53)	180,472 (47)	379,557
25	605,906 (100)	0 (0)	605,906
26	1,019,037 (78)	290,922 (22)	1,309,959
27	1,248,469 (88)	174,255 (12)	1,422,725
28	51,677,963 (13)	361,582,016 (87)	413,259,979
29	173,644 (4)	4,041,286 (96)	4,214,930
30	2,742,552 (78)	786,031 (22)	3,528,583
31	80,274 (100)	0 (0)	80,274
32	103,335 (1)	18,035,124 (99)	18,138,459
33	1,472,571 (93)	107,585 (7)	1,580,156
34	7,386,188 (85)	1,332,636 (15)	8,718,824
35	11,879,873 (98)	271,740 (2)	12,151,613
36	27,283,111 (69)	12,288,730 (31)	39,571,841
37	8,161,110 (62)	5,068,141 (38)	13,229,251
38	959,990 (94)	57,542 (6)	1,017,532
39	1,727,729 (49)	1,838,095 (51)	3,565,824
40	40,709 (100)	0 (0)	40,709
42	4,066,556 (100)	0 (0)	4,066,556
47	0 (0)	34,321 (100)	34,321

(continued)

TABLE 7 (continued)

SIC	Waste Volume (gals/yr)[b]		
	Recycled offsite	Recycled onsite	Total recycled
49	4,552,807 (56)	3,560,738 (44)	8,113,545
50	16,272 (5)	295,541 (95)	311,813
51	373,143 (100)	0 (0)	373,143
73	8,087 (0)	3,723,766 (100)	3,731,853
76	6,785 (100)	0 (0)	6,785
78	102,224 (54)	86,478 (46)	188,702
80	0 (0)	3,328 (100)	3,328
82	45,978 (100)	0 (0)	45,978
89	5,225 (100)	0 (0)	5,225
95	0 (0)	12,523 (100)	12,523
97	122,744 (60)	82,029 (40)	204,773
99	19,322 (34)	36,813 (66)	56,135
Total	128,549,584 (23)	424,001,745 (77)	552,551,329

[a]Source: Versar, 1985.

[b]Note: Volumes in gals/yr. Numbers in parentheses indicate percentages.

5. Facility A Case Study

Facility Description

Plant A was founded in 1981 and the Warwick, RI facility opened for business in June 1985. The company is involved in the development of a reclaimable product from metal plating waste baths and etchant dumps. Wastes received at the plant are pretreated to adjust pH and/or remove cyanides, and then converted to elemental metals or oxide sludges which are sold to smelting operations in Europe for precious metal recovery. The facility is currently operating under the precious metal recovery exemption of RCRA since sludge product is sold for its precious metal content (Ni, Cu, and Au). The facility's Part B TSD permit is currently being reviewed by Rhode Island DEM and approval is anticipated shortly.

The treatment/recovery facility is located in a light industrial section of the city of Warwick, RI. The recently constructed 30,000 sf facility houses administrative offices, a full laboratory, tank truck and tote (300 gal containers) unloading facilities, temporary storage (24 hr) for incoming wastes (4-4,000 gal tanks), raw material storage (100,000 gal), reactors, clarifiers, and solids handling facilities. Solids generated by the process are recovered and dewatered on plate and frame filter presses. The dewatered sludge is currently dried onsite although, during the site testing, offsite drying was employed.

Waste Sources

Facility A processes concentrated dumps from the metal plating and printed circuit board industries. These concentrated dumps include alkaline etchants, acid plating baths and electroless plating cyanide baths. Most of these wastes fall into the following four categories which provide a logical basis for segregation at Facility A:

- Acidic metals solutions;

- Alkaline metals etchant solutions;

- Cyanides; and

- Chelated metals solutions.

32

Average metals concentration in the process feedstock (incoming wastes) are approximately 12 g/L (12,000 ppm).

Waste Management

The unit operations employed to detoxify the wastes and recover metals include sodium hypochlorite oxidation of cyanides, pH adjustment, sodium metabisulfite reduction, sodium borohydride reduction, sedimentation, rapid sand filtration, dewatering (plate and frame filter press), and ion exchange columns for effluent polishing. Figure 3 shows the process schematic for the facility.

Waste Handling and Storage--

Incoming wastes or feedstock are transferred from 4,000 gallon tank trucks or totes (300 gal capacity containers) in the receiving bay, which is a fully enclosed multi-lined concrete-epoxy construction facility with 4,400 gallon capacity to contain spills. In order to minimize human error, four special fittings are provided in the receiving bay to handle each of the four wastes described above. When the waste is received, samples are taken for screening purposes to be compared to the waste anticipated from the delivering facility. While the samples are being screened the tank truck load is stored in one of four dedicated temporary storage tanks (less than 24 hours). If the screening results check, the waste is transferred to short-term raw material storage tanks where it is segregated by the following metals groupings as well as the four waste categories described earlier:

1. Cu, Ni, and other precious metals

2. Zn, Cd

3. Sn, Pb

4. Fe, Cr

If the screening results do not check properly with the contracted waste specifications then a full set of analyses are performed and the waste is classified prior to transferring the waste from temporary storage to raw material storage.

Facility A has approximately 100,000 gallons of segregated raw storage capacity. Six tanks provide 35,000 gallons storage for acid wastes, four tanks provide 25,000 gallons storage for alkaline wastes, five tanks provide 25,000 gallons storage for cyanide wastes, and three tanks provide 15,000 gallons storage for chelated wastes. The storage tanks are fixed roof tanks of concrete polyethylene lined construction. All piping is CPVC or butt-welded polyethylene construction. All storage tanks are vented to a building exhaust system equipped with a caustic scrubber.

Batch Reactor Tanks--

Three batch reactors are used at Facility A to process wastes for recovery (precipitation) of precious metals. One 15 cubic meter (3,963 gals) reactor is used for cyanide waste treatment and an identical reactor is

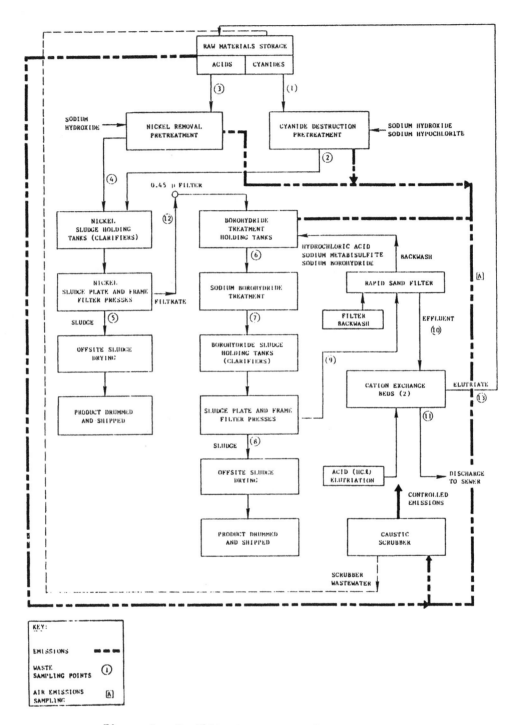

Figure 3.   Facility A process schematic.

dedicated to metal waste treatment. A larger 30 cubic meter (7,926 gallons) reactor is used for treatment of both acidic and alkaline wastes. The reactors are of polyethylene-lined concrete fixed roof tanks fitted with small manways, agitators, and large probes for measuring pH and oxidation-reduction potential. In addition, the reactors are equipped with chemical feed parts for automatic metering of chemical reagent. All reactor emissions are evacuated to the central exhaust system which is controlled by a caustic scrubber.

The chelated waste will be handled separately from other waste streams to avoid recombination of the chelating compounds with other metal ions. Chelated waste will be pretreated in approximately 4,000 gallon batches to further reduce the chelated metal complex or tie-up the chelate. For example, Facility A uses lime treatment to break EDTA complexes to form calcium EDTA salts. Chelates based upon quadrols and citrates can be reduced with sodium borohydride. The treatment technique applied is highly dependent on the metal ion complexed, the chelating agents used, metal ion concentration, and pH of the concentrate. Facility A initially conducts bench scale tests to determine optimum treatment methods for each chelated waste stream and uses the prescribed technique each time that the waste is received. Once pretreatment is completed, the waste is further treated with sodium borohydride as discussed below for acid/alkaline wastes.

The cyanide waste reactor generally serves as a pretreatment step to the sodium borohydride step. Cyanide wastes are treated by alkaline chlorination in the cyanide reactor. Batch sizes similar to the chelated wastes (approximately 3,200 gallons) are pretreated to reduce CN concentrations to less than 1 mg/L. Depending upon CN concentrations, batch reaction time ranges from 3 to 12 hours. The reaction steps proceed in the following order:

1. Adjustment of pH to 11 with sodium hydroxide.

2. Sodium hypochlorite (NaOCl) addition; the reaction is controlled by maintaining oxidation-reduction potential (ORP) at +400 mv.

3. Cyanide levels checked periodically until CN level is less than 1 mg/L.

4. Ferrous sulfate ($FeSO_4$) added to remove surplus chlorine (Cl_2).

5. Pretreated waste is pumped to the acid/alkali reactor for further treatment.

Most process batches at Facility A are eventually transferred to the acid/alkali reactor for sodium borohydride treatment. Batch sizes in the acid/alkali reactor are approximately 6,340 gallons (24 m^3). Reaction times are generally 2 to 2-1/2 hours for normal acid/alkaline wastes and 3 to 3-1/2 for chelated wastes. The reaction steps begin with pH adjustment which is usually accomplished in part by combining acid wastes with alkaline wastes in the reactor. The treatment steps are:

1.   Adjustment of pH to 6 with sodium hydroxide or sulfuric acid.

2.   Sodium metabisulfite addition (200-300 lbs per 24 cubic meter batch) for pre-reduction.

3.   Sodium borohydride addition (10 percent solution, 100 L per batch) with ORP and/or batch color monitored. When batch turns black, ORP is approximately -100 to -400 mv and reaction is complete.

During the above treatment significant quantities of hydrogen (H_2) gas may be liberated by foaming which takes place in the reactor. When the reaction is complete, the reactor contents are pumped to clarifiers, which are used as holding tanks prior to sludge dewatering.

Clarifiers (Sludge Tanks)--
     Facility A employs three 30 cubic meter rectangular sedimentation tanks for holding the reaction products from the reactor tanks while they are being fed to the plate and frame filter press. The clarifiers were originally designed to separate the reaction products (solids) from the aqueous supernatants (clarifier overflow). However, since the original precipitation reaction design was modified to borohydride reduction, the clarifiers are no longer required because the filter presses can easily dewater the entire batch volume. This is partly due to the large particle size of the agglomerated sludge resulting from borohydride treatment.

Plate and Frame Filter Press--
     Due to the plant modifications discussed above, the entire contents of the sludge holding tanks (clarifiers) are pumped to one of two plate and frame filter presses. Each press is capable of processing approximately 800 liters of sludge per hour. Metal sludge produced on the filter press should be in excess of 50 percent solids and 25 percent metal. From this point the sludge was previously shipped offsite for drying to reduce the moisture content from approximately 50 percent to 30 percent water. Since the completion of the testing program, Facility A has installed onsite infrared sludge dryers.

Rapid Sand Filter--
     The filtrate from the sodium borohydride plate and frame filter press is fed to the rapid sand filter. A single media rapid sand filter, rated at 10 cubic m/hr, provides some additional effluent polishing in the event of a sludge filter press failure and serves mainly to protect the cation exchange columns.

Cartridge Filtration (Prefilter)--
     A cartridge filter, rated at 10 m^3/hr, is used for polishing the discharge from the high pH (nickel pretreatment) plate and frame filter press filtrate prior to the sodium borohydride treatment step. The filter employs a design pore opening of 0.45μm.

Cation Exchange Columns--
     Two cation exchange columns in series serve as final polishing steps for the plant effluent prior to discharge to the City of Warwick, RI sewer system. The cation exchange columns are periodically elutriated with HCl which generates an acidic metals waste which is recycled to the acid/alkalai reactors for treatment.

Waste Characterization/Process Monitoring

Waste screening analyses are performed in an onsite laboratory as discussed earlier in the process description. Key process parameters which are monitored near the end of each batch include:

- pH

- ORP

- Cyanide concentration

- Metals concentration (total and dissolved)

- Temperature

- Color (visual)

- Moisture content (percent by weight)

Periodic monitoring of all permitted discharge limits are also conducted. In addition, the facility is equipped with cyanide alarms to warn of airborne cyanide concentrations buildup within the facility.

PROCESS TESTING AND ANALYTICAL RESULTS

On December 11 and 12, 1985, field studies were conducted to evaluate sodium borohydride waste treatment/reduction processes at Facility A. As the residuals (dried sludge) from the processes are sold for their precious metal content, this technology significantly reduces wastes which would otherwise be landfilled.

Test Deviations and Changes

Facility A plant process modifications required several minor alterations to the originally proposed test program. Under the original test program, samples were to be collected around unit process and from two process batches which would then be combined to make a single sodium borohydride process sludge sample. A process change (i.e., adding a sodium hydroxide precipitation pretreatment step) created an additional process reactor and sludge product to be sampled. In addition, this change dramatically increased the length of time (number of batches) necessary to fill the sodium borohydride sludge filter press. Thus, while the test plan originally called for sampling all streams from a single process batch in one 8-hour shift, the new process mode would have required nearly a week of sampling to obtain single batch data.

It was noted that because all of the batches come from the same feedstock tanks, sampling from different batches was presumed to be fairly representative. Thus, it was agreed that the investigators should spend an additional day sampling to collect as much data from a single batch as possible. However, it was also agreed that data from separate batches would be acceptable to the program requirements. As a result, the Facility A sampling program collected data from four process batches.

Results

In order to assess sodium borohydride as a viable waste treatment sludge reduction alternative, its effectiveness in meeting effluent requirements and obtaining low sludge volumes was evaluated. The parameters of interest are: trace metals, TOC/TOX, cyanides, and hexavalent chromium. Each parameter was examined for reduction efficiency, sludge content, and regulatory compliance. Other parameters of interest examined were air emissions, the effectiveness of the cyanide destruction system and the metals content in the sodium hydroxide sludge (Sample Point 5).

Trace Metals--

Reduction is defined as the gaining of electrons by an atom, an ion, or an element thereby reducing its positive valence. The success of its metal reduction is highly dependent on the mixing, residence time, and other process conditions such as: pH, temperature, concentration, and reaction kinetics. The purpose of the trace metals analysis is to evaluate sodium borohydride's effectiveness in the reduction of a mixed metal influent to a low volume, high density sludge. The streams of interest are: the borohydride reactor influent (Sampling Point 6), the borohydride reactor effluent (Sampling Point 7), and the borohydride sludge (Sampling Point 8).

It was initially proposed that a mass balance would be developed across the whole borohydride reduction process through the sludge filter press and its effluent. However, due to apparent variations in batch compositions and problems with the borohydride sludge press operation, it was necessary to use the results from Batch 3 (85-12-1009) to assess sodium borohydride effectiveness. The data from Batch 3 were particularly useful since effluent samples were collected and filtered after treatment which in effect simulated solids removal achieved in the sludge filter press.

Processing for Batch 3 took place in the 24 cubic meter sodium borohydride finishing reactor (acid/alkali) at ambient temperature and atmospheric pressure. The total metals loading in the batch reactor influent was 7.25 kg per batch, of which over 83 percent was divalent copper. The theoretical level of sodium borohydride (SBH) required for the total reduction of all metals was 8 kg (58 liters of a stabilized aqueous solution of 1.2 percent SBH and 4.1 percent caustic soda). The results presented in Table 8 summarize plant operations during testing. The actual sodium borohydride solution usage was 9.8 kg (70 liters of 1.2 percent SBH). This represents an actual/theoretical SBH addition ratio of 1.2, which falls well within the range of 1.0 to 1.5 reported in literature. Excess SBH is normally required due to nonoptimum reaction conditions and side reactions with other species such as aldehydes, ketones, nitrates, peroxides, and persulfates.

TABLE 8.   SODIUM BOROHYDRIDE FINISHING REACTOR PROCESS DATA

Treatment	pH	ORP[a] (mv)	Comment
-	10.5	-250	Start
Add 216 liters HCL	5.5	150	Lower pH
Add 200 lb NaHSO$_3$	5.2	130	Stabilizer (prereductant)
Add 60 liters NaBH$_4$	8.0	-770	Reduction
Add 25 liters Plexon	7.8	-770	SO$_2$ suppressant
Add 10 liters NaBH$_4$	8.4	-830	Finish

[a]Oxidation reduction potential

Sodium borohydride has a reducing capacity of 8 electrons/mole or an equivalent weight of 4.75 g/molar electron and a standard electrochemical potential of -1.25 volts (Lindsay and Hackman, 1985). Table 9 compares SBH reactor influent data with both filtered and nonfiltered effluent data for 8 selected metals. As expected, the nonfiltered effluent data demonstrate little or no reduction due to the effluent solution becoming resaturated during analysis for total metals. The filtered sample shows reduction with efficiencies ranging from 16.1 to 99.8 percent. This wide range of reduction efficiencies is likely a result of the concentration and chemical potential (activity) of each of the metallic ions contained in the solute.

Table 10 summarizes the reduction efficiencies of each of the selected metals in the filtered sample as a function of concentration and electrochemical potential. Analysis of the results show that the concentration of the metallic ion is often the determinant in reduction efficiency. However, given equivalent concentrations, the ability of a metallic ion to achieve equilibrium may be measured by its standard free energy or chemical potential. An example of this behavior was exhibited by lead and chrome which, under test conditions, have similar concentrations but divergent electrochemical potentials. The resultant 40 percent drop in reduction efficiency for trivalent chrome as compared to lead may be directly attributed to the greater quantity of free energy (approximately six times) required for chrome to achieve elemental form.

Analysis of filtered effluent showed that overall, of the 7.25 kg of mixed metals, approximately 6.91 kg were reduced to elemental form. This represents an overall reduction efficiency of 95.4 percent of total mixed metals. The remainder of the metals influent loading (0.337 kg), of which 70 percent was calcium, was of sufficient quality that given efficient post treatment, effluent limitations should be achieved.

A second objective in assessing sodium borohydride as a viable waste treatment alterative is the ability to form a low volume, high density sludge. An earlier study on hazardous sludge reduction (Aldrich, 1983) reported that substitution of SBH treatment for lime treatment of mixed metal wastewaters can result in a 68 percent sludge reduction. In addition, it has been reported (Heleba, May 84) that SBH reduction sludges typically contain 80 percent or more metals. These results compare favorably with the metals content of hydroxide-lime sludge which generally contains less than 20 percent metals. Since Facility A utilized both SBH and sodium hydroxide reduction, a trace metals analysis was conducted to determine sludge loading characteristics in each case.

The sludge samples collected from the nickel/cyanide sludge plate and frame filter press (Sample Point 5) were analyzed for 17 trace metals. The analytical results for these sludge samples (on a dry weight basis) are summarized in Table 11. The feed to the filter press consisted of the entire contents of the clarifier holding Batch 1 (85-12-1007) and Batch 2 (85-12-1008). These batches in turn consisted of the effluent from the cyanide reactor (Sample Point 2) and the effluent from the nickel pretreatment reactor (Sample Point 7) which included filtrate from the nickel/cyanide press.

TABLE 9.   SODIUM BOROHYDRIDE FINISHING REACTOR TRACE METALS
            CONCENTRATIONS AND REMOVAL EFFICIENCIES

Element	Reactor influent (mg/L)	Reactor effluent (mg/L) unfiltered	Reactor effluent (mg/L) filtered[a]	Percent of removal
Ag	24.0	6.2	0.06	99.7
Au	5.7	4.76	0.15	97.0
Cd	0.015	0.01	0.01	b
Cr	0.031	0.03	0.026	16.1
Cu	237.0	207.0	0.47	99.8
Ni	0.96	0.902	0.422	56.0
Pb	0.32	0.31	0.14	56.2
Zn	5.10	4.76	0.079	98.4

[a]Filtered onsite at Plant A's Lab. In addition, a blank DI water sample was
filtered at the Plant's Lab onsite as a QC measure. Results for that sample
showed less than detection limits in all cases.

[b]Unable to obtain adequate precision.

TABLE 10.   REDUCTION EFFICIENCIES AS A FUNCTION OF CONCENTRATION AND
ELECTROCHEMICAL POTENTIALS IN BATCH 3

Element	Concentration (mg/L)	Electrochemical potential	Reduction efficiency (%)
Cu	237.0	0.3402	99.8
Ag	24.0	0.7996	99.75
Au	5.7	1.42	97.37
Zn	5.1	-0.7628	98.45
Ni	0.96	-0.23	56.04
Pb	0.32	-0.1263	56.25
Cr	0.31	-0.74	16.13
Cd	0.15	-0.4026	-

TABLE 11.   FACILITY A SLUDGE CHARACTERIZATION RESULTS

Element	Dry weight concentration (percent)		EP Toxicity results (mg/L)		EP Toxicity standards[a] (mg/L)
	Ni/CN sludge	SBH sludge	Ni/CN sludge	SBH sludge	
Ag	0.019	0.017	0.03	0.06	5.0
As	0.004	0.017	0.04	0.05	5.0
Au	0.134	0.328	---	---	---
Ba	0.001	0.001	0.224	0.163	100.0
Ca	0.205	0.089	---	---	---
Cd	0.008	0.001	0.589	0.016	1.0
Cr	0.029	0.003	0.294	0.032	5.0
Cu	11.000	5.250	---	---	---
Fe	0.720	0.049	---	---	---
Mg	0.042	0.003	0.0018	0.0022	0.2
Ni	19.400	0.293	---	---	---
Pb	0.130	0.046	4.6	0.03	5.0
Rh	0.865	0.050	---	---	---
Se	0.003	0.001	0.04	0.04	1.0
Sn	0.305	0.115	---	---	---
Tl	---	---	---	---	---
Zn	2.840	0.361	---	---	---
Total	35.705	6.624	---	---	---

[a]Source U.S. Environmental Protection Agency
 Federal Register   V. 45 No. 98 98:   33122 May 14, 1980

Analysis of the sodium hydroxide (NaOH) and SBH sludges yielded total metals dry weight fractions of 35.7 and 6.6 percent, respectively. Neither sludge results supported Facility A's claim of 60-70 percent metals on a dry basis. While the SBH sludge result was significantly below performance expectations, the exact cause of these results was not discernable. Possible explanations include: 1) a possible process upset; 2) sampling error; or 3) analytical error. It seems most probable that a process upset was responsible for these results, since blinding of the sludge press did occur on the SBH press. Based on the results of other case studies on SBH reduction conducted under this program, it is reasonable to assume that these results are not representative, since typical sludge metals contents should be greater than 70 percent.

Table 11 also presents Facility A sludge EP toxicity leachate results for both the nickel/cyanide and sodium borohydride reactor sludges. The results of the tests clearly show that for Facility A influent metals concentrations, the sodium borohydride sludge is fairly stable in that its leachate characteristics are below EP toxicity limits for all metals. However, note that the waste is still classified as F006 hazardous waste.

An additional objective of this study was to demonstrate that Plant A's treatment system was in compliance with final effluent limits. Once the filtrate leaves the SBH plate and frame filter press it is fed to a single media rapid sand filter (Sample Point 10) to provide some additional effluent polishing. Final polishing is performed in cation exchange columns (Sample Point 11) prior to discharge to the city of Warwick sewer system. Samples were collected at Sample Point 11 for final effluent from Batches 1 and 3. Table 12 presents metals concentration results in comparison with local pretreatment effluent limits. In both samples, the quality of the effluent was inadequate to meet local effluent limits.

In an effort to remedy this problem, Facility A revised its waste processing sequence in the following manner:

- Incoming noncyanide wastes nave been adjusted to pH 7.5 with sodium or magnesium hydroxide.

- Copper, zinc (refinery brass) and trace amounts of cadmium and silver are reduced and pumped into filter press one.

- The filtrate from filter press one is transferred back to the reaction tank and the pH is adjusted to 11.5 with NaOH, thus precipitating any nonreduced metals as the hydroxide.

- Nickel and trace amounts of heavy metals are reduced and collected in filter press two.

- The filtrate from filter press two is transferred to the SBH finishing reactor and precious metals such as gold, platinum, palladium, and chrome are reduced.

TABLE 12. PLANT A LOCAL PRETREATMENT EFFLUENT COMPLIANCE DATA

| Element | Total metals concentrations (mg/L) | | |
	Batch 1	Batch 3	Effluent limit[a]
Ag	0.05[b]	0.05[b]	0.03
As	0.62[c]	0.29[c]	0.01
Cd	0.01	0.01	0.05
Cr	0.29[c]	0.03	0.2
Cu	0.146	1.82[c]	0.4
Ni	0.767[c]	0.861[c]	0.5
Pb	0.14	0.1	0.15
Se	0.2	0.2	1.0
Zn	0.98[c]	0.05	0.5

[a]City of Warwick, RI effluent limits.

[b]Unable to obtain great enough precision.

[c]In excess of pretreatment effluent limits.

- The reduced metals precipitates are then removed through bag and micron filters prior to final effluent polishing in the cation exchange columns.

In addition, Facility A has instituted the use of a QC holding tank following the cation exchange columns to prevent column breakthrough. In this manner Facility A is able to prevent any discharge to the sewer that might exceed effluent limits. Since testing was completed with these revisions of the process sequence, Facility A's effluent has been tested by the local sewer district authority on several different occasions. Since implementation of these changes, Facility A's effluent quality (based on local sewer district authority sampling results) has improved considerably and is now consistently meeting sewer authority guidelines.

Organic Indicator Results--
Total Organic Carbon (TOC) and Total Organic Halide (TOX) samples were extensively collected and analyzed for Batches 2, 3, and 4. Total organic carbon results are summarized in Table 13 and total organic halide results are presented in Table 14. As expected, both sample runs display little, if any, reduction in organic concentrations after being processed through the SBH finishing reactor. This phenomena may be due to the fact that the sludge results do show organic constituents being concentrated in the sludge. Total TOC/TOX concentration in the nickel/cyanide sludge and the SBH sludge were 2.6 and 4.0 percent, respectively. Thus, while the sludge results show some concentration of organics, influent and effluent results showed little or no removal.

Cyanide Results--
Total cyanide samples were collected and analyzed for Batches 3 and 4. The unforeseen presence of distillable organics in the Facility A process streams may have shown a positive bias in the test results, particularly in sample points 2 though 13. Results obtained for the influent and effluent samples collected at the CN destruction reactor are considered semi-quantitative due to marginal QC recoveries. Based on the QC tests and the relative unimportance of the presence of cyanide in determining the effectiveness of a sodium borohydride reduction, system results for cyanide were not incorporated into the body of this report. The analysis should be considered incomplete. Refer to the QC section for a detailed explanation of analytical quality and control results.

Hexavalent Chromium--
Due to the complex nature of Facility A's process flow streams, the investigators were unable to obtain any acceptable hexavalent chromium results. The presence of complex organics and strong reductants were apparently the cause of the poor hexavalent chromium precision and accuracy results. Discussion of these QA/QC problems are presented later in Section 9.

Process Emissions--
In addition to assessing wastewater effluent characteristics, the testing program was designed to evaluate uncontrolled process air emissions. Table 15 summarizes the results of grab sample and integrated sample analysis of process reactor exhaust ducts based on Draeger tube analysis. The emission

TABLE 13. SUMMARY OF TOTAL ORGANIC CARBON RESULTS (mg/L)
FACILITY A – WARWICK, RI

Sample point No.	Description	Batch 1 (85-12-1007)	Batch 2 (85-12-1008)	Batch 3 (85-12-1009)	Batch 4 (85-12-1010)
1	CN Reactor Influent				
2	CN Reactor Effluent				
3	NI Pretreatment Influent				633
4	NI Pretreatment Effluent			729	587
5	NI/CN Sludge		25,100		
12	NI/CN Press Filtrate [a]				
6	Borohydride Influent			468	
7	Borohydride Effluent [c]			511/528 [b]	
8	Borohydride Sludge [d]		38,600		
9	Borohydride Filtrate [e]			632	
10	Sand Filter Effluent			722	
11	Ion Exchange Resin Effluent			618	
13	Ion Exchange Resin Elutriate				

[a] Sampled after Filter Press and 0.45 μm filter.
[b] Includes sludge from batches 2 and 3.
[c] In addition to effluent sample, a separate sample was collected and filtered onsite at the Plant lab. As a QA measure, a DI water blank was also filtered by Plant A.
[d] Includes sludge from batches 1 and 2.
[e] Sampled after sand filter and 0.45 μm filter.

TABLE 14.   SUMMARY OF TOTAL ORGANIC HALIDE RESULTS (mg/L)
FACILITY A - WARWICK, RI

No.	Sample point Description	Batch 1 (85-12-1007)	Batch 2 (85-12-1008)	Batch 3 (85-12-1009)	Batch 4 (85-12-1010)
1	CN Reactor Influent				
2	CN Reactor Effluent				34
3	NI Pretreatment Influent				13
4	NI Pretreatment Effluent			6.5	
5	NI/CN Sludge		1,000		
12	NI/CN Press Filtrate[a]				
6	Borohydride Influent			23	
7	Borohydride Effluent[c]			17/13[b]	
8	Borohydride Sludge[d]		1,500		
9	Borohydride Filtrate[e]			26	
10	Sand Filter Effluent			39	
11	Ion Exchange Resin Effluent			20	
13	Ion Exchange Resin Elutriate				

[a] Sampled after Filter Press and 0.45 μm filter.
[b] Includes sludge from batches 2 and 3.
[c] In addition to effluent sample, a separate sample was collected and filtered onsite at the Plant lab.
As a QA measure, a DI water blank was also filtered by Plant A.
[d] Includes sludge from batches 1 and 2.
[e] Sampled after sand filter and 0.45 μm filter.

TABLE 15.   SUMMARY OF DRAEGER TUBE ANALYSIS RESULTS FOR UNCONTROLLED
PROCESS AIR EMISSIONS[a]

| Parameter | Gas Concentrations (ppm or % as noted) | | Threshold limit value (TLV) short term exposure limit[b] |
	Grab Sample Results	Integrated Sample Results	
Hydrogen Cyanide	<2 ppm	<2 ppm	10 ppm[c]
Hydrogen	1.7 - 6.0[d] %	0.4 %	--
Sulfur Dioxide	<1 - 20 ppm	<1 ppm	5 ppm
Hydrogen Sulfide	<1 ppm	<1 ppm	15 ppm
Ammonia	<5 - 180 ppm	<5 ppm	35 ppm
Hydrochloric Acid	<1 ppm	2 ppm	5 ppm[c]

[a]Draeger detector tubes are compound-specific for the parameter indicated.
Accuracy is estimated at ± 5-20% of reading. Test conditions were as follows:

Flowrate = 3,600 afpm
Duct diameter 6 inches
Duct area = 28.274 in^2 or 0.196 ft^2
Volumetric flowrate at actual conditions = 0.196
ft^2 X 3,600 afpm = 706.86 acfm.

[b]Source: ACGIH 1985.

[c]Time weighted average value used in lieu of short term exposure limit.

[d]Five pump strokes were required (10 strokes standard) to reach saturation
concentration of 3%, thus extrapolated reading is 3.0% $\frac{(10)}{(5)}$ = 6.0%.

results given in Table 15 show a frequent presence of hydrochloric acid and
hydrogen gas accompanied by occasional presence of ammonia and sulfur
dioxide. Grab sample concentrations for ammonia and sulfur dioxide exceeded
adopted short term exposure limits (ACGIH, 1985) for these substances. One of
the hydrogen emissions grab sample results (6.0 percent) is significant since
this value is greater than the lower flammable limit for hydrogen
(4.0 percent). This is primarily due to hydrogen gas being evolved during SBH
treatment and is likely to be a function of the pH of the wastewater. This
problem may be eliminated through optimization of the treatment process and
should remain a design consideration for new applications.

ECONOMIC EVALUATION

    As previously discussed, a sodium metabisulfite/borohydride treatment
process was used at Facility A to reduce complex process solutions to metallic
form. A primary obstacle to the more widespread use of sodium borohydride has
been its high cost. Table 16 summarizes cost and performance of various
reduction chemistries as a function of pound of copper removed. Lime/ferrous
sulfate is the least expensive reducing agent, but will increase sludge
generation by at least 68 percent relative to sodium bisulfite/borohydride
(Aldrich, 1983). In addition, lime/ferrous sulfate sludge is difficult to
sell and refine due to its low metals content (5 percent metals) and high
gypsum content. An inability to sell the sludge product would result in a
RCRA permit violation since Facility A currently operates under a precious
metal recovery exemption. Therefore, sodium borohydride was the reductant of
choice since its precipitant will yield finely divided metals which are easily
recovered. Chemical costs for SBH reduction typically range from $6.80 to
$17.00 per pound of copper removed depending on the actual to theoretical
usage ratio (1.0 to 5.0).

    Initially, Facility A operated its sodium metabisulfite/borohydride
reduction reactor with plexon, an additional reducing agent. Plexon is a 12
weight percent dimethyldithiocarbamate solution used to reduce nickel and to
lower sulfur dioxide emissions. However, a relatively high cost of $8 per
liter rendered the entire reduction process impractical at $19.80/lb copper
removed and also decreased sludge loading characteristics. By deleting plexon
from the reduction reaction, Facility A was able to decrease chemical costs
63 percent to approximately $7.30 per pound of copper reduced.

    Further process optimizations at Facility A have included the 2-stage
reduction process prior to the SBH finishing reactor and the installation of
onsite infrared drying ovens. These changes have reportedly resulted in a
sludge product that is 90 percent solids and 60 percent precious metals on a
dry weight basis. In addition, chemical costs have decreased to approximately
$6.00 per pound of metal reduced since only a small volume of sodium
borohydride solution is required.

TABLE 16.   ECONOMIC COMPARISON: PRINTED CIRCUIT BOARD
            WASTEWATER TREATMENT

Treatment	lb chemical Per lb Cu reduced	Chemical cost[a] $/lb Cu reduced
$FeSO_4$ [b]	9 - 44	0.9 - 4.4
VenMet[TM] Solution[c]	5.7 - 14.2	6.8 - 17.0
Facility A[d]	21.7	19.8
Facility A[e]	13.9	7.27

[a]Chemical Costs   Source: Ventron Technical Brochure
    $0.1/lb $FeSO_4 \cdot 7H_2O$
    $0.44/lb 40% Sodium dimethyldithiocarbonate solution
    $2.40/lb VenMet Solution
    $0.25/lb Sodium Bisulfite

[b]Source: Ventron Technical Brochure

[c]Source: Ventron Technical Brochure, contains 12% by weight sodium
  Borohydride and 40% NaOH, as well as 3.2 - 8.0 lbs Sodium Bisulfite/lb
  Cu reduced

[d]Measured during testing, contains VenMet solution, sodium bisulfite, and a
  12% sodium dimethyldithiocarbonate solution (plexon)

[e]Revised system containing VenMet solution and sodium bisulfite

6. Facility B Case Study

Facility Description

Facility B is a manufacturer of printed circuit boards based in Santa Ana, California. Ten years ago Facility B began operating on a job shop or contract basis. The company currently employs 77 people, operating two shifts, 5-1/2 days/week. The facility has now been in their present location for 5 years and are planning an expansion of their operations. Printed circuit board production is approximately a half-million square feet/year, generating $7 million of gross sales.

Several years ago Facility B was discharging to the sewer as much as 40 lb/day of untreated chelated and particulate copper. In April 1984, legislation was introduced that would require printed circuit board manufacturers to limit their effluent streams to 2.7 ppm copper. In response to this pending legislation, Memtek Corporation in Woburn, Massachusetts was hired by Facility B to perform a study to determine an appropriate waste treatment system. As a result of this study, the facility installed a sodium borohydride reduction and membrane ultrafiltration waste treatment system in February 1983.

Facility B's wastewater discharge is permitted (Class I wastewater discharge permit) and sampled on a quarterly basis by the Orange County Sanitation District. If any of the effluent limitations are exceeded, sampling is performed more frequently and corrective actions are taken.

Waste Sources

The three main methods of printed circuit board production are the additive, subtractive, and semi-additive techniques. Additive techniques involve the production of printed circuit boards through electroless plating on unclad board materials. Subtractive involves the removal of large amounts of copper foil from clad board material to create the desired circuit pattern. The process used at Plant B is a hybrid between the two aforementioned methods called semi-additive. Figure 4 illustrates the Plant B process and Table 17 presents some of the chemicals used in various steps of the process. The waste stream of interest is metals containing wastewaters from the rinses following the etching and plating operations as well as production bath dumps. Consideration of the following general and specific process areas can assist in evaluating metals sources within the production process.

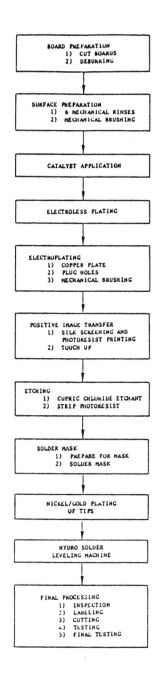

Figure 4.   Plant B process flow diagram.

TABLE 17.   CHEMICALS USED IN PLANT B'S PROCESS

Process step	Chemicals used
Catalyst Application	Hydrochloric Acid (HCl) Cadmium (Cd)

Electroless Plating	Copper Sulfate ($CuSO_4$)
	Sulfuric Acid (H_2SO_4)
	Ammonium Persulfate (NH_4, S_2O_8) –
	    used only occasionally to clean
	    electroless line

Electroplating	Copper Sulfate ($CuSO_4$) Sulfuric Acid (H_2SO_4)
Image Transfer	Caustic Soda, liquid and anhydrous (NaOH) – used as a developer
Etching	Sodium Chloride (NaCl) Sodium Chlorate ($NaClO_3$) Muriatic Acid (HCl)
Hydro Solder Leveling Machine	1,1,1-trichloroethane ($C_2Cl_3H_3$)

Drilling and Deburring--
    During board preparation, the boards are sawed into blanks slightly
larger than is needed for the final product to allow for tabs and board
finishing. After mechanical or chemical cleaning the typical double-sided or
multilayered board is drilled by numerically controlled high speed spindle
drills. The resulting holes and board edges are deburred by rotating brushes
to remove any loose particulate matter or rough edges detrimental to
subsequent chemical processing.

Electroless Copper Deposition, Rinsing, and Neutralization--
    After chemical cleaning and rinsing (to remove any dirt or surface oils),
the boards are catalyzed through the application of a thin layer of stannous
and palladium chloride. The stannous chloride layer is removed prior to
electroless plating by a mild fluoroboric acid solution (accelerator). This
removal exposes the palladium chloride ion which acts as a catalyst in the
subsequent electroless copper reaction. The electroless copper reaction
typically deposits a thin (25 to 85 micro-inch) layer of copper on the board
surface and in the drilled holes. This metal deposition provides electrical
contact between the surfaces and layers of the printed circuit board. After
electroless copper deposition, the boards are thoroughly rinsed and then
neutralized with a mild sulfuric acid solution.

Electrolytic Plating--
    In order to ensure a uniform electrical conductivity, i.e., no breaks or
voids in the copper layer, a 1 to 2 mil deposit of electrolytic copper is
deposited on the thin electroless layer. The general processing procedure is
to activate the board surface with hydrochloric acid (to remove any surface
contaminants), plate, clean/rinse, and replate. Acid copper plating baths
contain sulfuric acid, copper sulfate pentahydrate, organic brighteners, and
50 to 70 ppm of chloride ions. Deposition takes place through the reduction
of cupric ions by electrical current which flows through the cell from anode
(phosphorized copper bars) to cathode (plating surface).

Positive Image Transfer--
    Image transfer involves the production of a circuit pattern on a
metallized board surface with an ultraviolet light sensitive organic polymer.
In order to create a positive image, the photoresist is first applied directly
to the copper surface by a hot roll laminator or silk screening. Then a
stencil of the artwork is exposed to ultraviolet light while under vacuum to
produce the exact circuit pattern. Upon exposure, the photoresist surrounding
the polymerized circuit pattern becomes soluble. A caustic soda solution is
used to develop the photoresist and removes any nonpolymerized material. The
remaining nonsoluble photoresist, i.e., the circuit pattern, is now a chemical
inhibitor and acts as an etch resist.

Etching and Resist Strip--
    Cupric chloride is used as an inexpensive final etch process for boards
without metallic etch resists. Its main constituents are cuprous chloride,
sodium chloride, sodium chlorate, muriatic acid, and water. The overall
reaction is:

$$CuCl_2 + Cu \rightleftharpoons 2CuCl$$

This etching solution will remove all unwanted or excess copper from the board, that is not protected by the polymerized photoresist. Following the etch process, the boards are rinsed (an important source of metallic contamination), and then immersed in a high temperature, alkaline photoresist stripping tank containing butyl cellosolve acetate as the active ingredient.

Solder Mask--

A solder mask or resist is a polymer coating which is applied to a printed circuit board to prevent molten solder from adhering to preselected areas. Solder resists act as a protective coating, preventing harmful elements from degrading the circuits. In addition to physical protection, solder masks also serve as an electrical insulator. The resist liquid is applied through silk screen printing prior to curing, which fully crosslinks the resist polymers to achieve proper end use characteristics.

Gold/Nickel Microplating--

The section of the printed circuit board that contacts the main assembly is known as the gold edge connector. The edges are chamfered to allow easy insertion and are designed and manufactured for maximum conductive and corrosion resistance properties. Gold tends to form an intermetallic layer with copper (changing its properties) and is too ductile for most applications. Therefore, after a mild activation step, a 50 to 100 micro-inch layer of nickel is plated over the copper to act as a hardening agent and prevent the migration of copper molecules. Gold is then plated in a potassium cyanide bath containing organic brighteners to provide the final protective coating on the connector edges. Both nickel and gold baths are examples of electrolytic plating and are followed by rinses. The gold, however, is recovered directly in process due to its expense.

Hot Air Leveling--

The selective solder coating/hot air leveling process involves applying an eutectic solder coating onto the copper areas not covered by the solder mask. Prior to application, the gold edges are masked with tape to prevent solder adhesion, the board is precleaned and then thermally conditioned/ activated by a water soluble flux. As the board exits from the solder, it passes through two heated, horizontal air knives, producing a quality, selected deposit. After final rinsing and solvent cleaning using 1,1,1-trichloroethane (to remove any residual flux), the board goes to final inspection.

Waste Management

As discussed above, the main sources of metallic contamination to the wastewater stream emanate from the rinses following plating or etching operations. Copper contamination is confined to the rinses following copper chloride etching, electroless and electrolytic plating, and the activation baths on the electrolytic and microplate lines. Nickel contaminants are introduced solely from the rinse following the electrolytic nickel plating operation on the microplate line. Lead contamination is introduced into the waste stream in the rinsing and cleaning operation following hot air leveling.

The purpose of this case study is to evaluate metals containing hazardous sludge (RCRA code F006) minimization technologies. The process of interest is the semicontinuous, sodium borohydride ultrafiltration system. The primary operation is the precipitation of heavy metals through the use of a strong reducing agent (sodium borohydride) followed by liquids-solids separation through membrane ultrafiltration and sludge filtration. A brief description of the two technologies is presented below.

A description of the process used by Plant B in the reduction and precipitation of incoming complex wastes is represented in Figure 5. Process water from the plant (streams 1 and 2) is collected in a reservoir, combined with filtrate from the filter press (stream 6), and subsequently pumped into a chemical reaction tank for pretreatment. A level controller in the reservoir activates a sump pump that initiates the transfer of influent upon reaching the required volumes. Since the transfer rates are highly contingent upon a variable wastewater flow rate, all subsequent operations have been designed to perform in a semicontinuous or batch mode.

The wastes entering the chemical reaction tank (stream 3) are automatically adjusted for pH through the use of sodium hydroxide or sulfuric acid. Since many heavy metals are only soluble under acidic conditions, to facilitate the reaction, pH is maintained in the alkaline range (7 to 11). An ORP (oxidation reduction potential) controller automatically meters a dilute (12 percent) sodium borohydride solution to ensure complete reaction and optimize chemical consumption. The resultant reaction products flow by gravity into the concentration tank.

Sodium Borohydride Reduction--

Sodium borohydride (SBH) is a strong reducing agent and provides a simple and efficient method of metal precipitation and recovery. SBH is able to reduce metal contaminants to their elemental form which results in a low volume, high metal content sludge. In addition, the use of SBH promotes good settling characteristics which minimize the need for a flocculant.

An understanding of the chemistry associated with the use of SBH is helpful. The basic reduction reaction involves the donation of 8 electrons/molecule of SBH to an electron deficient metal cation. The following half-reaction occurs when SBH is added to an aqueous effluent:

$$NaBH_4 + 2H_2O \rightleftharpoons NaBO_2 + 8H^+ + 8e- \tag{1}$$

If this reaction takes place in the presence of metal cations, reduction occurs according to the following reaction:

$$8M^+ + 8e^- \rightleftharpoons 8M^o \tag{2}$$

If there are no inorganic or organic reducibles, hydrolysis takes place:

$$NaBH_4 + 2H_2O \rightleftharpoons NaBO_2 + 4H_2 \tag{3}$$

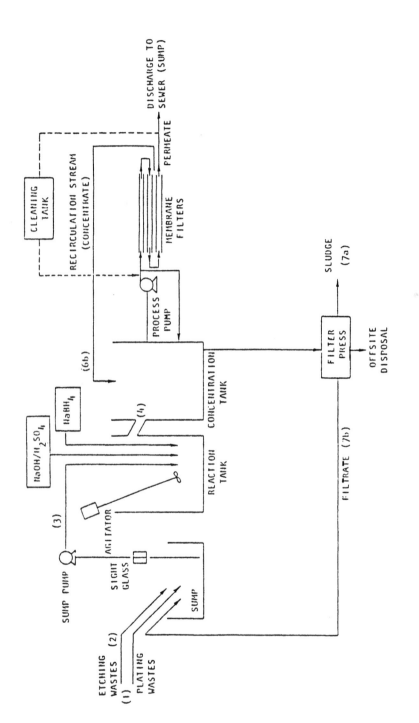

Figure 5.   Process schematic showing Plant B plating/etching waste treatment system.

Combining Equations 1 through 3 yields the overall reaction:

$$NaBH_4 + 2H_2O + BMX \rightleftharpoons NaBO_2 + 8M^O + 8HX \qquad (4)$$

where: M = metal (valence + 1), and

X = anion (chloride, carbonate, etc.).

This reaction is dependent on the following process conditions and operational parameters: pH, temperature, metal concentrations, the kinetics of competing reactions, agitation, residence time and the method of liquid-solid separation. The pH should be in the slightly alkaline range. Though reaction temperatures vary from one application to the other, it should be noted that reduction is usually rapid at ambient temperature.

Ultrafiltration--

Ultrafiltration (UF) is the second aspect of Plant B's system to be discussed. The UF process discriminates on the basis of molecular size, shape and flexibility. Suspended solids and large molecule colloidal solids (0.002 to 10 μm) can be filtered in this process. In conventional filtration, flow is perpendicular to the surface of the filter. In ultrafiltration, flow is in the direction parallel to the surface of the filter. A pressure drop of 35 psi is employed. A summary of the Facility B's ultrafiltration system design parameters is shown in Table 18.

The concentration tank acts as a repository for the solids generated in the sodium borohydride reaction. Concentration is achieved by forcing the permeate through the 0.1 micron pores of the ultrafiltration membrane. The suspended solids are rejected and returned to the concentration tank while the permeate is discharged to the sewer (stream 5). Multiple passes are employed to improve overall removal efficiency while a level controller prevents the pump from cavitating. On a regular basis (twice/day), the concentration tank is drained and its contents dewatered to a 20 to 40 percent solids in a filter press. The resultant filtrate is returned to the sump (stream 6), while the filter cake is disposed offsite (stream 7).

To prevent membrane blockage or damage, chemical cleaning by a sodium hypochlorite solution is performed daily. The cleaning solution is circulated throughout the ultrafiltration unit for approximately 20 minutes, dissolving any metals that have become trapped in the membrane. In the case of critical blockages, the system design permits isolation and easy replacement of any of the seven ultrafiltration modules.

Sludge Filter Press--

The mechanical sludge dewatering device in use at Facility B is a Delta Unifilter low-pressure filter press. Pressure filters of this type dewater sludge by pressurizing it and forcing the permeate out through a membrane. The Facility B filter press has an operational area of 7.5 ft^2 and a line pressure of 65 to 75 lb/ft^2. Sludges at Facility B are usually dewatered to 20 to 40 percent solids. The final solids concentration depends on the length of time the sludge remains in contact with the filter and the operating pressures applied to the sludge.

TABLE 18.   PLANT B ULTRAFILTRATION SYSTEM SPECIFICATIONS[a]

Parameters	Design
Solids content of influent (recycle)	1-2%
Pressure drop	35 psi
Waste throughput	2 gpm
Filter area (total)	15 ft^2
Cycle time	batch
Pore size	0.1 μm
Flux	200 gal/ft^2/day
Tube diameter	1 in.
Number of tubes	10

[a]Per module, 6 to 10 modules/system.

Source: Memtek Ultrafiltration Systems

PROCESS TESTING AND ANALYTICAL RESULTS

Process Testing

The objective of the sampling program was to evaluate the effectiveness of the waste reduction technology utilized by Facility B. The effectiveness was measured in terms of volume reduction of hazardous waste streams and minimization of other releases to the environment. The waste stream parameters analyzed under the sampling and analysis program included total metals (copper, nickel, lead, and zinc), EP toxicity metals (sludge filter cake only), total organic carbon (TOC) and total organic halides (TOX). Each parameter is evaluated in a comprehensive mass balance which focuses on three main streams of interest. The main streams of interest consist of the combined influent into the reaction tank (stream 3), the permeate effluent from the ultrafiltration unit (stream 5), and the sludge filter cake from the filter press (stream 7a).

On January 7, 1986, sampling was performed on the Facility B waste treatment system beginning at 9:15 a.m. As stated in the QA Plan, samples were taken on an hourly basis or as available. Table 19 summarizes the reaction tank sampling times and indicates deviations from the hourly sampling schedule in several instances. The pH of streams 3, 4, and 5 were measured using field instrumentation.

Flow measurement readings were also monitored from plant instrumentation during the initial stages of testing. However, it was soon obvious that the flow metering equipment at Plant B was malfunctioning and other means for estimating flowrate were employed.

One source of flowrate estimates was obtained by contacting the Orange County Sewer Authority for recent data on Facility B wastewater flowrates. This source indicated an average flowrate of 22,000 gpd was measured during quarterly monitoring. A second estimate of actual flowrates was developed from the throughput of the SBH/ultrafiltration wastewater feed pump (1-1/2 hp). The feed pump, which is manufactured by Gould Inc., is rated for 27 to 35 gpm with a 4-inch suction and a 3-inch discharge. However the pump operated on an intermittent basis (approximately 75 percent of the time), thus the effective flowrate was approximately 1,215 to 1,575 gph for 16 hours or 19,440 to 25,200 gpd. The two methods for estimating flowrate provided a good check against each other. Based on this information, the measured value of 22,000 gpd was used.

Analytical Results

The data summarized in Table 20 show influent and effluent stream concentrations for metals of interest in this study. Influent copper concentrations during testing (786 ppm) significantly exceeded normal levels indicated by the facility (150 to 200 ppm). This apparent anomaly may have been due to process abnormalities during testing, although the plant personnel did not mention any unusual occurrences. During GCA's sampling, the influent waste stream was observed to turn a dark shade of blue for a short period of time, indicating a temporary increase in metallic copper content. Although

TABLE 19.   PROCESS OBSERVATIONS AND MEASUREMENTS

| Time | Reaction tank | | |
	pH[a]	ORP[b] (mv)	Comments
9:15	14.42	-440	Stream 1 not flowing enough to collect a sample.
9:45	14.51	-570	
9:55	14.51	-565	
10:10	14.48	-835	Membrane cleaning began at 10:00. Used 75 gallons of NaOCl.
10:20	14.47	-863	Membrane cleaning ends.
10:45	13.67	-838	Reaction tank contents are blue.
10:55	8.60	+050	Reaction tank contents are black.
11:10	8.58	-052	Reaction tank contents are green/brown.
11:20	8.53	+090	
11:45	11.83	-461	Influent to sump stopped at 11:30 (lunch break).
12:10	11.95	-578	
12:20	11.67	-277	
12:30	14.34	-450	
12:45	14.42	-459	
1:05	14.24	-257	
1:20	14.23	-339	
1:45	8.91	+058	
2:00	14.48	-286	
2:35	14.51	-319	
2:45	14.52	-297	
3:00	14.05	-252	
3:10	8.92	+054	

[a] pH readings from Plant metering equipment are apparently high, readings greater than 14.0 are probably not valid.

[b] ORP = Oxidation Reduction Potential.

TABLE 20.   SUMMARY OF TEST DATA FOR SODIUM BOROHYDRIDE
            TREATMENT AT PLANT B

Parameter	Influent waste (Stream 3) (mg/L)	Effluent wastewater (Stream 5) (mg/L)	Sludge[a] (Stream 7) (μg/g)
Total organic carbon	40.0	36.1	184.8
Total organic halide	1.76	1.75	–
Select trace metals:			
Cu	786.0	1.42	780,000
Ni	0.055	0.03	58.7
Pb	0.57	0.10	300
Zn	3.86	0.028	1,430
EP toxic metals			
As	–	–	0.03
Ba	–	–	0.522
Cd	–	–	0.002
Cr	–	–	0.003
Pb	–	–	1.8
Hg	–	–	0.0002
Se	–	–	0.04
Ag	–	–	0.56

[a]Results given on a dry weight basis for sludge.

this fluctuation does not affect the daily mass balance, drawing any annual
estimates from the concentration data could be misleading. A second
observation drawn from the raw data is the 78 percent (dry weight) of copper
in the filter cake, which compares favorably with the vendor supplied data of
80 to 95 percent (Linsey and Hackman, 1985). Both total organic carbon and
total organic halide were virtually unaffected by the reduction process with
only 9.75 and 0.57 percent losses to the sludge stream.

Trace Metal Results--
The objective of the trace metal analysis was to evaluate both the
efficiency of reaction and the removal efficiencies observed for the sodium
boronydride/ultrafiltration treatment system at Plant B. The first evaluation
utilized a process mass balance approach to determine actual and theoretical
reagent requirements and calculate the effectiveness of the sodium borohydride
reagent in reducing the metals of interest. The second evaluation involved an
assessment of influent and effluent concentrations and a comparison of these
with local and Federal effluent limitations to determine process viability.

In developing a mass balance for metals contained in the entering and
exiting wastestreams, it was necessary to make the following assumptions:

1. Flowrate was constant at 22,000 gpd;

2. Wastewater flow to Stream 7 was small compared to Streams 3 and 5;
    and

3. The influent flowrate (Stream 3) was approximately equal to the
    effluent flowrate (Stream 5) at 22,000 gpd.

These assumptions were necessary due to difficulties encountered during
testing with Plant B's wastewater flow metering equipment.

Metals concentration data were used in conjunction with waste stream
flowrate information to develop the mass balance results presented in
Table 21. Based on these data the sodium borohydride treatment system at
Plant B showed high removal/recovery of copper, zinc and lead while showing
reduced effectiveness for nickel. During testing, the total metals loading to
the SBH reactor and ultrafiltration system was approximately 145 lb/day of
which over 99 percent was divalent copper (Cu^{+2}). The theoretical
requirement for the total reduction of all metals present in the SBH reactor
influent stream was estimated to be 175.5 lbs of sodium borohydride
(15 gallons of an alkaline solution containing 12 percent by weight of SBH).
The actual quantity of solution consumed during the reduction reaction was
148.6 lbs (12.7 gallons of solution). This consumption rate represents an
actual to theoretical SBH solution addition ratio of 0.84. Previous case
studies (Heleba, 1984) have shown that SBH requirements are more typically 1
to 1.5 times the theoretical requirement. It is speculated that SBH reduction
operated below stoichiometric requirements due to an absence of competing
reactants (e.g., aldehydes) and relatively high metals concentration found in
the reactor influent.

TABLE 21. MASS BALANCE OF TRACE METALS RESULTS FOR
SODIUM BOROHYDRIDE TREATMENT AT PLANT B

Total metals analyte	Stream 3 Influent waste (lb/day)	Stream 5 Effluent wastewater (lb/day)	Stream 7 Sludge (lb/day)	Percent recovery (%)
Cu	144.1524	0.2604	143.892	99.82
Ni	0.0101	0.0055	0.0046	45.54
Pb	0.1045	0.0183	0.0862	82.49
Zn	0.7079	0.0051	0.7028	99.28

Metals concentrations in the effluent stream were used to determine the effectiveness of the SBH reduction system in both meeting effluent guidelines and minimizing releases to the environment. Table 22 describes effluent performance characteristics in terms of reduction efficiencies and effluent compliance. Analysis of these characteristics show that copper was reduced most efficiently at 99.82 percent, while nickel removal was the least efficient at 45.54 percent. The wide disparity in removal efficiencies seems to be mainly a function of concentration (higher concentrations are removed more efficiently), but the chemical potentials (quantity of free energy required for an ionic species to obtain equilibrium) may also have been a factor. Approximately 144.7 lbs of total metals were reduced to elemental form by the SBH ultrafiltration system, representing a total reaction efficiency of 99.8 percent. Overall the quality of the effluent produced by the SBH treatment system was quite good. Metals currently discharged to the sewer are now meeting stringent local and Federal EPA pretreatment standards. Previously Facility B was unable to meet County standards using a batch filtration system which frequently failed, unintentionally discharging precipitated copper (Lopez, September 1984). However, since the installation of the SBH ultrafiltration system, Facility B's effluent has been consistently below discharge requirements based upon sampling and analysis by the Orange County Sewer Authority.

An additional criteria in the assessment of the SBH/ultrafiltration treatment system is the characterization of the sludge filter cake. As Facility B currently ships the sludge product offsite for land disposal, the E.P. Toxicity leachate characteristics of the SBH sludge have been evaluated. Analysis of the raw data shows that the sludge product contains greater than 78 percent (dry weight) elemental copper. This concentration in conjunction with 1,430 g/g of reduced zinc combine to form a product called refinery brass. This intermetallic product can be easily recovered by a smelter, thus eliminating the need for land disposal and limiting any liabilities thereof. However, if the sludge product must be landfilled, the results in Table 23 show that leachate resulting from the sludge dry filter cake is within Federal EPA guidelines. However, it is noted that this resultant sludge would still be classified RCRA waste code F006 under current Federal regulations.

Total Organic Carbon/Total Organic Halide--
The objective of the total organic carbon (TOC) and total organic halide (TOX) analysis was to determine total organic loadings within the SBH reactor system. As mentioned previously SBH is an extremely efficient reductant and will reduce organic, as well as inorganic species. Often the presence of nonmetallic compounds such as aldehydes, ketones, nitrates, peroxides, and persulfates will reduce reactor efficiency and consume up to twice the theoretical quantity of SBH solution required. However, the relatively low concentrations of organics in the SBH reactor influent (presented in Table 24) showed little reduction of nonmetallic species. For example, TOC reduction was only 9.75 percent while TOX reduction was 0.57 percent. It is believed that low concentrations of these organic species contributed to the overall success of the SBH reactor system with respect to metals.

TABLE 22. PLANT B EFFLUENT PERFORMANCE CHARACTERISTICS

Metal[a]	Effluent[b] Concentration (mg/L)	Effluent[b] Loading (lbs/day)	County[c] standards (lbs/day)	Federal[d] limitations (mg/L)
Cu	1.42	0.2604	0.50	3.72
Ni	0.030	0.0055	0.70	3.51
Pb	0.10	0.0183	0.10	0.67
Zn	0.028	0.0051	0.70	2.64

[a]Measured as total trace metals method 3050.

[b]Permeate from Memtek ultrafiltration unit discharged to sewer.

[c]Orange County Sanitation District.

[d]Daily maximum (mg/L) for electroplating point source effluent limitations U.S. EPA, Federal Register U.77, No. 169:38477, August 31, 1982.

TABLE 23. EP TOXICITY LEACHATE RESULTS FOR PLANT B
SODIUM BOROHYDRIDE SLUDGE

Element	Concentration (mg/L)	EPA standards[a] (mg/L)
Arsenic	0.03	5.0
Barium	0.522	100.0
Cadmium	0.002	1.0
Chromium	0.003	5.0
Lead	1.8	5.0
Mercury	0.0002	0.2
Selemium	0.04	1.0
Silver	0.56	5.0

[a]U.S. Environmental Protection Agency,
Federal Register, Vol. 45, No. 98,
98:33122, May 14, 1980.

TABLE 24. ORGANIC LOADING RESULTS FOR FACILITY B SBH REACTOR SYSTEM

Stream ID	Description	TOC (ppm)	TOX (ppm)
3	SBH reactor system influent	40.0	1.756
6a	SBH reactor system effluent	36.1	1.746
7a	SBH filter cake	184.8	-

ECONOMIC AND ENVIRONMENTAL EVALUATION

Economic Evaluation

One of the objectives of this study was to evaluate the economics of the waste minimization technology. For this case study the economics of the SBH treatment system tested are presented along with lime-ferrous sulfate conventional technology which would be used in its place.

Central to any discussion or comparison between sodium borohydride and lime-ferrous sulfate is the actual to theoretical chemical usage ratio. In sodium borohydride applications, the high unit cost of sodium borohydride solution ($2.7/lb) versus ferrous sulfate ($0.11/lb) necessitates the careful control of chemical usage. Parameters of importance are: presence of nonpriority reducibles, pH adjustment, good mixing and settling conditions, adequate reaction time, and liquid/solid separation. These site-specific factors combined with effluent limitations and total treatment and disposal costs, can significantly affect the economics of employing sodium borohydride treatment technology.

The SBH application at Facility B is fairly uncharacteristic in that they utilize a cupric chloride instead of an ammonical etchant. This is significant in that cupric chloride contains very few complexants, which would interfere with the SBH reduction reaction, lowering reaction efficiency and driving up treatment chemical costs.

Complexing and chelating agent applications in the electronic components industry are typically for the suspension of metals in plating or etching solutions. Major sources of complexing agents are alkaline (ammoniacal chloride) and ammonium persulfate etchants. Borohydride may react with these other compounds (i.e., ammonia) in the wastewaters, thus reducing its availability for metal ions. Therefore, most electronics components users find that a large excess of borohydride is frequently required to ensure rapid and complete metals reduction.

A detailed cost analysis for both sodium borohydride and lime/ferrous sulfate (LFS) technologies is presented in Table 25. The Alliance test data clearly established that SBH treatment is superior to LFS treatment in this application, when capital costs are held constant. Chemical costs for the SBH treatment system were three times greater than LFS, however the more significant sludge disposal costs for SBH reduction are shown to be 93.5 percent less. As a result of these factors, SBH treatment was able to reduce overall operating expenses by 48 percent (in comparison to LFS treatment), while decreasing sludge production at the same time. In addition, Facility B will soon be practicing sodium borohydride sludge reclamation onsite. This will not only further reduce operating expenses, but also potentially lower liabilities associated with hazardous waste land disposal.

TABLE 25.   PLANT B ANNUAL TREATMENT AND DISPOSAL COSTS
FOR SODIUM BOROHYDRIDE AND LIME FERROUS
SULFATE PRECIPITATION TECHNOLOGIES

Basis	Unit cost ($)	Sodium borohydride treatment system cost ($/yr)	Lime ferrous treatment system cost ($/yr)
Chemical costs			
SBH solution	2.7/lb	100,298	-
Sodium hydroxide	0.32/gal	12,500	-
Ferrous sulfate	0.11/lb	-	35,888
Hydrated lime	50.0/ton	-	495
Total chemical cost		112,798	36,383
Disposal costs			
Sludge disposal[a]	200/ton	13,278[b]	205,560[c]
Annual costs			
Total annual cost		126,076	241,943
Cost/lb metal reduced		3.5	6.7

[a]35 percent total solids

[b]78 percent metal in the solids, 144.69 lb/day total metals loading in
sludge.

[c]5 percent metal in the solids, 144.69 lb/day total metals loading in
sludge.

Environmental Evaluation

Results indicate that the use of SBH is an effective means for metallic waste precipitation and solid waste management. As stated previously, SBH application is very site specific and the presence of oxidizing agents such as complexants can increase chemical demand by as much as 50 to 100 percent. However, its cost-effective performance in achieving discharge limits and reducing hazardous waste at Facility B makes it a practical alternative to comparative waste treatment technologies.

7. Facility C Case Study

Facility Description

This facility manufactures electronic computing equipment including logic, memory and semiconductor devices, multilayer ceramics, circuit packaging, intermediate processors and printers. Approximately 11,000 persons are employed at this particular location.

Waste Sources

Two of the major processes in which hazardous waste streams are generated are the manufacture of semiconductors and the manufacture of printed circuit boards. As mentioned in the previous section, the manufacture of these two products can involve the use of organic solvents both for the cleaning of surfaces and the developing and stripping of photosensitive resists. The photoresists are used to form either a positive or a negative image of the circuit pattern on the substrate chip or circuit board. After application of the photoresist to the substrate material, a mask of the circuit pattern is placed over the board or chip and the surface is exposed to light. Since in this case a negative photoresist material is used, the resist polymerizes upon exposure to light, while the resist that is covered by the photomask does not.

Following this exposure to light, developer solvent is used to remove the resist material which has not been stabilized. The developer solvent used at this facility is methyl chloroform (1,1,1-trichloroethane). Subsequent to developing the resist, the exposed areas of the substrate material are etched and/or metal plated. Once this has been done, the resist has served its purpose and it can be "stripped" from the surface. Either acids or organic solvents may be used for photoresist stripping. At this facility methylene chloride is used to strip photoresist from electronic panels. However, the spent solvent from this operation is handled separately and will not be addressed further in this report.

Waste Characteristics and Quantities--

Waste solvent streams will vary in composition both according to whether they were used for stripping or developing, and whether they were used in the manufacture of circuit boards or semiconductors. Several different organic solvents are used at this facility, including:

Methylene Chloride - Resist stripping of Electronic Panels

Methyl Chloroform - Resist Developing of Electronic Panels and
                       Substrate Chips

Freon - Surface Cleaning and Developing of Substrate Chips

Perchloroethylene - Surface Cleaning of Electronic Panels

The major difference in the waste solvents from resist stripping and
resist developing is that resist stripping solvents will contain polymerized
resist while resist developing solvents will contain unpolymerized resist.
Unpolymerized resists may polymerize if they are heated to a certain
temperature, and therefore waste containing these materials may have to be
heated differently than wastes containing already polymerized resists. In
both of the waste streams, resists are present as dissolved solids at maximum
concentrations of 1 percent by weight. The exact concentration of dissolved
solids in the solvent will depend on the volume of work processed in the
solvent.

After developing or stripping, the work piece is generally rinsed in
water to remove the residual solvent. This results in a waste aqueous stream
contaminated with solvent. Gravity settling is employed to separate the
solvent and water fractions directly after the developing or stripping
operation, but some residual amount of water may remain in the solvent
fraction. The water fraction is sent to wastewater treatment.

Most of the solvent waste streams are kept segregated to facilitate
recovery, but the developing process for substrate chips involves the use of
both methyl chloroform and Freon. Consequently, the spent solvent stream from
this process apparently contains a mixture of both of the solvents, 90 percent
Freon and 10 percent methyl chloroform.

Another solvent that is used at the facility is perchloroethylene. It is
used for precleaning the surface of electronic panels to remove dirt, oil or
grease which may have been deposited during previous manufacturing operation.
The spent solvent from this cleaning operation is handled separately and will
not be addressed further in this report.

Waste Management

The purpose of this study is to evaluate onsite methods of recovering
and/or recycling hazardous wastes. At Facility C the primary
recovery/recycling operation are the recovery and reuse of solvents by
distillation or evaporation. Several types of equipment are used to recover
spent solvents at this facility. Box distillation units are used to recover
methylene chloride and perchloroethylene, flash evaporators are used to
recover methyl chloroform and a distillation column is used to recover Freon.
These unit operations are described below.

Flash Evaporation of Methyl Chloroform--

Two flash evaporation units are used to recover spent methyl chloroform (MCF) from several different resist developers. The spent solvent from each of these developing areas is first treated to remove water and then pumped to a waste solvent collection tank. From this waste collection tank, the solvent is pumped to the flash evaporation units, where the contaminants are removed, and the recovered solvent is pumped to a clean solvent storage tank. Virgin methyl chloroform is added to the recovered methyl chloroform to replenish corrosion inhibitors, and the mixture is returned to the developers. A schematic of this system is shown in Figure 6.

Figure 7 shows the major components of the flash evaporation units and Table 26 indicates normal operating parameters. This type of a unit is used, instead of a conventional still, because of the presence of unpolymerized photoresist in the methyl chloroform. As mentioned above, the photoresist will polymerize when subjected to high temperatures such as those required to boil MCF at atmospheric pressure. If the resist polymerizes onto the heating coils of the still, the operation of the unit would be adversely affected. In flash evaporation, the solvent is preheated to approximately 100°F and then enters a "flash" chamber where a vacuum of 20.5" Hg causes the preheated liquid methyl chloroform to vaporize. At atmospheric pressure, the MCF must be heated to 165°F for boiling to occur.

In the flash chamber, a certain fraction of the MCF vaporizes, and the other fraction, containing the contaminants, remains in liquid form. The vapor passes through a condenser and is recovered at an average rate of 600 gallons per hour. The fraction that does not vaporize collects in the bottom of the chamber and is recirculated through the heat exchanger at a rate of 490 gallons per minute. A certain amount of this liquid is bled off every ten minutes and pumped to the still bottom storage tank.

The flash evaporators are fed from a 15,000 gallon spent solvent collection tank, and recovered solvent is collected in an adjacent 15,000 gallon tank. When the level of spent solvent reaches 12 or 13 thousand gallons, and the level of recovered solvent is down to 2 or 3 thousand gallons, the evaporators are turned on and operated until these quantities are reversed. This amounts to recovering approximately 10,000 gallons, which at a recovery rate of 600 gallons per hour for each unit, requires the operation of both units for approximately 8 hours per day.

The recovery efficiency of the flash evaporators ranges between approximately 95 and 98 percent. The recovered solvent will usually contain less than 20 ppm of dissolved solids and the still bottoms will contain up to 5 percent solids. Still bottoms are sent offsite for further recovery of methyl chloroform.

Distillation Column for Recovery of Freon--

A batch distillation system consisting of a distillation column equipped with stainless steel mesh packing is used to recover a waste stream containing spent Freon and methyl chloroform. As mentioned above, Freon and methyl chloroform are used respectively to preclean substrate chips and develop photoresist applied to the chips. The distillation column recovers the Freon

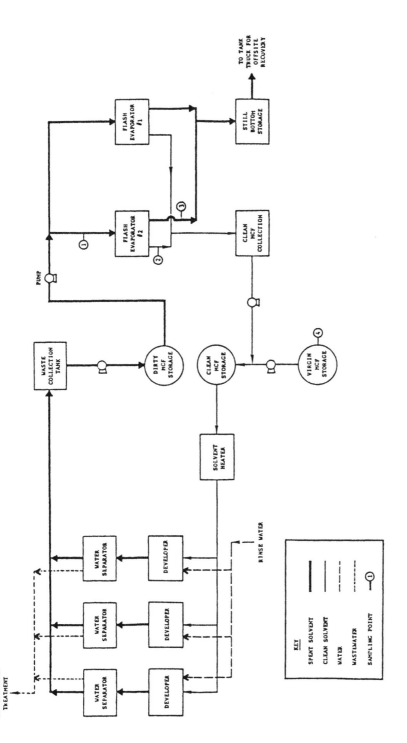

Figure 6. Schematic of methyl chloroform recovery system.

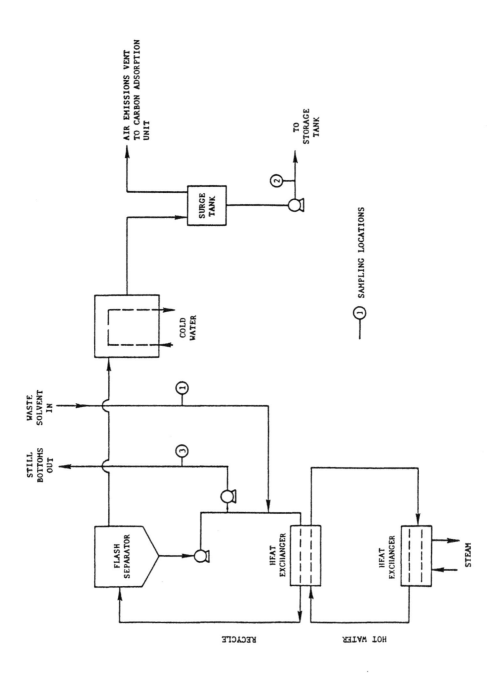

Figure 7.   Schematic of flash evaporator.

TABLE 26. METHYL CHLOROFORM FLASH EVAPORATION SYSTEM OPERATING INFORMATION
NORMAL DISTILLATION UNIT OPERATION

Operating Parameters	Normal Conditions
1. Solvent	
Recovery Rate	600 gph
Boiling Temperature	99°F (clean) 113°F (14.5% non-volatiles)
Recycle Rate	490 gpm
Distillate Temperature	70°F
Separator Pressure	20.5" Hg Vacuum
2. Chilled Water	
Flow	46 gpm
Inlet Temperature	47°F (normal)
Outlet Temperature	77°F (normal)
3. Steam	
Flow	800 lbs/hr
Pressure at Still Supply	30 psig
4. Hot Water	
Inlet Temperature	140°F (max.)
Outlet Temperature	133°F
Recycle Rate	400 gpm
5. Electrical Requirements	30 hp

as a distillate and generates still bottoms containing methyl chloroform and other contaminants such as dissolved solids. The recovered Freon is reused onsite, and the methyl chloroform still bottoms are either further distilled onsite using box stills or sent offsite for recovery.

A schematic of the system is shown in Figure 8 and normal operating parameters are presented in Table 27. Spent solvent from several developing stations is collected in a holding tank (1,500 gallon) located adjacent to the distillation column. From the holding tank, the solvent is fed into a 1,450 gallon batch pot equipped with a reboiler to heat the solvent to approximately 133°F. The feed rate from the holding tank to the batch tank is such that the level in the reboiler will be kept at 70 percent full. The system is rated to handle 1,200 gallons per hour.

The distillation column itself is 28 feet high and 12 inches in diameter. The vaporized solvent rises up the column with the Freon passing through and being recovered in a shell and tube vent condenser, while the MCF condenses on the packing and falls back into the batch tank. The average rate of recovery of Freon is 33 gallons per hour. The Freon which is recovered is collected in a tank and returned to be used for developing and precleaning.

This system is operated virtually 24 hours per day 7 days per week. As distillation proceeds, the concentration of methyl chloroform in the batch tank increases from an initial concentration of 10 percent in the waste feed to approximately 80 percent after 11 to 13 days. The fraction of MCF in the batch tank is determined by taking a sample of the contents and determining its specific gravity. When the specific gravity is 1.368, the MCF content is approximately 80 percent, and the still bottoms are removed. This removal occurs once every 10 or 11 days.

The quantity of Freon recovered in 1984 was 2,310,000 pounds, and the quantity of Freon/methyl chloroform still bottoms sent offsite for recovery was 1,355,000 pounds. The purity of the recovered Freon is generally greater than 98 percent. Analyses of the waste streams associated with the distillation column are presented in a subsequent section.

PROCESS TESTING AND ANALYTICAL RESULTS

Process Testing

Testing of the recovery processes at this facility took place on February 4 and 5 of 1986. The objective of the testing was to demonstrate the effectiveness of the distillation/evaporation processes in recovering solvents for onsite reuse. In order to do this, samples of influent, recovered product, and distillation bottoms were collected and analyzed for volatile and semivolatile organics and total solids to determine the fraction of both solvent and contaminant. Samples of the virgin solvent were also analyzed for comparison with the recovered solvent.

Flash Evaporator--
On the day of testing, one flash evaporator was not operating because of a broken pump. Operation of the other unit was initiated at approximately 7:40 a.m.. This unit was operated continuously for the length of the testing

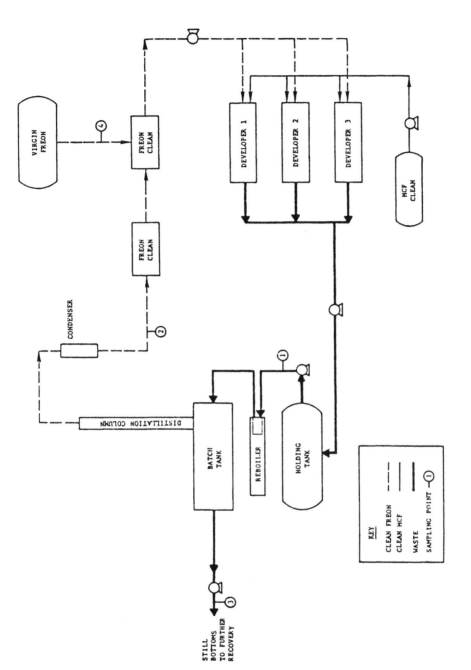

Figure 8. Schematic of methyl chloroform/Freon recovery system.

TABLE 27. METHYL CHLOROFORM/FREON BATCH DISTILLATION SYSTEM
OPERATING INFORMATION

Operating Parameters	Normal Conditions
1. Solvent	
Recovery Rate	33 gal/hr
Boiling Temperature	133°F
Distillate Temperature	75°F
2. Cooling Water	
Flow	72 gal/min
Temperature	85°F
3. Steam	
Flow	470 lbs/hr
4. Electrical Requirements	30 hp

period which ended at approximately 3:30 p.m. the same day. The unit
continued to operate after testing had been completed. The total amount of
waste that had been fed to the unit between 7:40 a.m. and 3:30 p.m. was
4,885 gallons (as measured by the influent totalizer). The total quantity
recovered in this period was 4,709 gallons. The average rate of recovery was
601 gallons per hour, which is equivalent to normal operating conditions.
Other operating parameters were monitored during the testing period and were
found to be within ±10 percent of normal. Solvent recycle rate was 475 gpm,
boiling temperature was 108°F, and separator pressure was 19 in Hg vacuum.

Distillation Column--
      The distillation column was sampled over a period of six hours. Samples
of the waste influent, distillate and distillation bottoms were collected at
2 hour intervals. One variation between the actual sampling and that proposed
in the QA plan was that the three distillation bottom samples were composited
in the field to provide one sample for analysis instead of collecting three
individual samples. All other samples were also composited except for those
that were to be analyzed for volatile compounds.

      At the time of the testing, the most recent still bottom removal had been
approximately 5 days beforehand. Since still bottoms are generally removed
once every 11 to 13 days, when the methyl chloroform content reaches
80 percent, it would be expected that the fraction of Freon at the time of
testing was still fairly high.

Process Testing Results

      The methodology used to assess solvent recovery process performance
included the following three parameters: comparing the recovered product with
virgin material; comparing the relative fractions of contaminant in the waste
influent with those in the recovered product; and comparing the residues
(still bottoms) from the recovery process with the waste influent. The
performance of the flash evaporator and the distillation column with respect
to these parameters is discussed below.

Flash Evaporation--
      Table 28 presents analytical results for process and waste streams
associated with the flash evaporation unit. In comparing the composition of
the recovered product with the virgin material it is evident that two are
virtually identical, having the same fraction of methyl chloroform and a
difference of only 0.8 mg/kg of solids and 16 mg/kg semivolatiles.

      The second method of evaluating efficiency is to compare the composition
of the waste influent and the recovered product. The data in Table 28
indicate that the concentration of dissolved solids in the recovered product
was less than 99 percent of the dissolved solids concentration in the waste
influent. In addition, 96 percent of the volume of spent solvent influent was
recovered by evaporation.

      One can also assess efficiency based on the composition and quantity of
the still bottoms. Evaporation of the spent solvent stream resulted in a
still bottom stream of approximately one thirtieth of the volume and two

TABLE 28.   FLASH EVAPORATION RESULTS

Parameter	Influent	Recovered Product	Still Bottoms	Virgin Methyl Chloroform
Volatile Organics (% w/w)				
— Freon TF	<0.1	<0.1	<0.1	<0.1
— Methyl Chloroform	99.9	99.9	92	99.9
— Other	<0.1	<0.1	<0.1	<0.1
Semivolatiles[a] (mg/kg)	82	145	1155	129
Total Dissolved Solids (mg/kg)	460	3.0	78,000	2.2
EP Toxic Metals (mg/l)				
— Arsenic			0.03	
— Barium			0.121	
— Cadmium			0.002	
— Chromium			0.003	
— Lead			0.03	
— Mercury			0.0009	
— Selenium			0.04	
— Silver			0.01	
Stream Volume (gal) Processed	4,885	4,709	176[b]	

[a]Unidentified compounds.

[b]By subtracting recovered product from influent.

hundred times the solids concentration of the waste influent. Nonetheless, the bottoms still contained 92 percent methyl chloroform and only about 8 percent contaminant. It would be possible to further evaporate these still bottoms and recover more of the methyl chloroform, however the product recovered may not be as pure. In addition, the solids concentration of the still bottoms could increase to a level that would affect the operation of the evaporator. As the solids concentration rises, the boiling temperature rises, and polymerization of the solids could occur.

The analysis of the still bottoms for EP Toxic concentrations of metals proved negative. This test is relatively meaningless for these still bottoms since they still contain a high concentration of solvent. As a result they will be sent offsite for further recovery instead of being land disposed. The still bottoms from further recovery,however, may contain higher levels of metals which would be a factor in final disposal.

Distillation Column--

Table 29 contains results of analyses on waste and process streams associated with the distillation column. A comparison of the data on the recovered Freon and the virgin Freon indicates that a high degree of recovery was achieved, with the exception of its methyl chloroform content. Whereas the virgin Freon contained non-detectable levels of MCF, the recovered product contained approximately 1 percent. Solids concentrations in the recovered product, however, were only 0.06 mg/kg versus 0.13 mg/kg in the virgin material.

A comparison of the waste influent and recovered product indicates that a 95 percent reduction in solids is achieved, and that the Freon content is raised from 96 percent to 99 percent.

One point that must be noted is that the still bottoms are not continuously being removed from the batch tank. They are only removed when the methyl chloroform concentration reaches 80 percent. The data presented in Table 29 indicate that the still bottoms product are not yet ready to be removed. A sample was taken primarily to indicate the relative fractions of the two solvents at the time of the site visit.

ECONOMIC AND ENVIRONMENTAL EVALUATION

Economic Evaluation

In this section, detailed cost estimates are presented based on 1986 cost factors (unit costs). These cost factors are presented in Table 30.

The cost estimates themselves are presented in Tables 31 and 32. They are based on the quantities of waste solvent and still bottoms that were generated in 1984. Most of the cost factors are self-explanatory, but those pertaining to offsite residue disposal (or management) require some explanation. Based on discussions with several waste disposal firms (not those that currently handle this facilities' wastes), a per gallon cost was derived both for the offsite recovery of still bottoms generated by onsite

TABLE 29.   DISTILLATION COLUMN RESULTS

Parameter	Influent	Recovered Product	Still Bottoms	Virgin Freon TF
Volatile Organics (% w/w)				
- Freon TF	96	99	52	99.9
- Methyl Chloroform	3.9	0.9	47.9	0.1
- Other	0.1	0.1	0.1	0.1
Semivolatiles[a] (mg/kg)	86	71	161	67
Total Dissolved Solids (mg/kg)	1.2	0.06	27	0.13
Stream Volume (gal) Processed	300[b]	198[b]		

[a]unidentified compounds.

[b]Estimated based on average rates.

TABLE 30.   UNIT COSTS USED IN CALCULATING PROCESS ECONOMICS

Electricity[a]	$0.05/kw-hr
Steam[b]	$0.96/1000 lb
Cooling Water[c]	$0.25/1000 gal
Operating Labor	$15/hr
Engineering	10% of equipment cost
Other Capital Costs	10% of equipment cost
Annualized Capital	Based on 10 years and 10% interest
Methylene Chloride[d]	$3.90/gallon
Methyl Chloroform[d]	$4.50/gallon
Freon 113[d]	$11.92/gallon
Offsite Recovery of Spent Solvent (Residue Disposal)	$0.25/gallon credit
Offsite Recovery of Still Bottom (Residue Disposal)	No charge

[a]Department of Energy, Energy Information Administration. National Average, December 1986.

[b]Peters, M.S. and K.D. Timmerheus. Plant Design and Economics for Chemical. McGraw-Hill p. 389 (updated using Chemical Engineering cost factors).

[c]EPA-625/5-85/016. Environmental Pollution Control Alternatives Reducing Water Pollution Control Costs in the Electroplating Industry. September 1985.

[d]Chemical Marketing Reporter. May 5, 1986.

TABLE 31.   ESTIMATED COST OF METHYL CHLOROFORM RECOVERY

Cost Item	Onsite Recovery Quantity	Onsite Recovery ($) Cost[a]	Offsite Recovery Quantity	Offsite Recovery ($) Cost[a]
Capital Cost				
- Equipment	2	320,000	--	--
- Engineering	--	32,000	--	--
- Other	--	32,000	--	--
TOTAL CAPITAL		384,000		--
Annual O&M				
- Electricity	30 kw	8,400	--	--
- Steam	800 lbs/hr	4,300	--	--
- Cooling Water	9,600 gal/hr	13,400	--	--
- Labor	2,800 hrs/yr	42,000	--	--
- Maintenance	--	38,400	--	--
- Residue Disposal	129,000 gallons	0	3,619,000 gallons	(904,700)
TOTAL O&M		106,500		(904,700)
Annual Costs				
- Annualized Capital	--	68,000	--	--
- Annual O&M	--	106,500	--	--
- Solvent Cost	129,000 gallons	329,800	3,619,000 gallons	10,911,300
TOTAL COST		504,300		10,006,600

[a] 1986 dollars.

TABLE 32. ESTIMATED COST OF RECOVERY OF FREON/METHYL CHLOROFORM

Cost Item	Onsite Recovery		Offsite Recovery	
	Quantity	($) Cost[a]	Quantity	($) Cost[a]
Capital Cost				
- Equipment	1	130,000	0	0
- Engineering	--	13,000	--	--
- Other	--	13,000	--	--
TOTAL CAPITAL		156,000		0
Annual O&M				
- Electricity	1.5 kw	600	--	--
- Steam	470 lbs/hr	3,800	--	--
- Cooling Water	4,320 gal/hr	9,100	--	--
- Labor	2,800 hrs/yr	42,000	--	--
- Maintenance	--	15,600	--	--
- Residue Disposal	119,500 gallons	0	295,800 gallons	(74,000)
TOTAL O&M		71,100		(74,000)
Annual Costs				
- Annualized Capital	--	12,600	--	--
- Annual O&M	--	71,100	--	(74,000)
- Solvent Cost	23,900 gallons	190,900	176,300 gallons	1,408,000
TOTAL COST		274,600		1,334,000

[a]1986 dollars.

distillation, and for offsite recovery of spent solvent in the hypothetical case that onsite distillation was not practiced. In the former case, it was assumed that the waste management facility would accept the still bottoms, containing 5-10 percent solids, for no charge, but no credit. In the latter case, where the solvent would contain less than 1 percent solids, the waste management facility would give the generating facility a 25 cent per gallon credit for each of the solvent types.

Another assumption was made to determine the cost that the waste management facility would charge to the industrial facility to buy back the recovered solvent. This cost was assumed to be two-thirds of the current price of the virgin solvent. Therefore, the solvent costs presented in Tables 31 and 32 were calculated using per pound costs that are two-thirds of what is shown in Table 30.

A discussion of specific factors related to the recovery of the solvent waste streams is presented below.

Methyl Chloroform--
The estimated costs associated with onsite recovery of methyl chloroform by flash evaporation are displayed on the left side of Table 31, and those associated with offsite recovery are displayed on the right-hand side of the same table. The onsite recovery costs are based on the use of an APV Paraflash evaporator capable of handling a 600 gallon per hour feed rate. The FOB cost of each of the two units is $160,000 and this includes a plate heat exchanger, a vapor liquid cyclone separator, shell and tube main condenser, shell and tube vent condenser, hot water set for indirect steam heating, a set of pumps and instrumentation for automatic operation [APV Crepaco, 1986]. The equipment would also be fully preassembled on a skid with manual valves and piping. The capital cost estimate also includes engineering and "other" costs because at this facility modifications are generally made to equipment to meet site-specific conditions.

O&M costs are based on the operation of each of the units, for 8 hours each day 350 days per year. Labor costs were estimated only for operation of the units. It was assumed that one person would be assigned to monitor operation of the two units, 8 hours each day 350 days per year.

In looking at the costs, it is evident that the major cost is associated with the purchase of virgin solvent. Since the purchase cost of virgin methyl chloroform is $4.50 per gallon versus approximately 5 cents per gallon (of recovered solvent) to operate the flash evaporator, it certainly makes economic sense to recover the solvent onsite. When the solvent is sent offsite for recovery it can be bought back at two-thirds the price of virgin solvent, but this is still a cost of $3 per gallon.

The annual savings resulting from onsite recovery is greater than $9 million. One of the major reasons for the tremendous savings is that the amount of spent solvent generated is so large. At a smaller facility, savings would not be quite so impressive.

Freon/Methyl Chloroform--
     Costs associated with the recovery of Freon are presented in Table 32.
The equipment cost is based on the use of a 1,200 gallon per day APV Batch
distillation system. The $130,000 FOB price includes a 28-foot distillation
column with metal mesh packing, a 1,450 gallon batch pot, a U bundle reboiler,
U bundle condenser, shell and tube vent condenser, bottom and top product
pumps, bottom and top product shell and tube coolers, and instrumentation for
automatic operation [APV Crepaco, 1986]. The unit would be preassembled and
include all valves and piping.

     O&M costs for this system are based on operating this unit 24 hours per
day, 350 days per year. This unit is operated continuously as long as spent
solvent is available for input. Labor costs are those for operating the
unit. Since the unit is equipped with instrumentation to allow for automatic
operation, it is not necessary to have someone monitor operation 24 hours per
day. Instead, an estimate of 8 hours per day was used as a maximum amount of
operating labor required.

     Onsite recovery of Freon results in a cost savings of approximately
$1 million per year compared to sending the waste solvent offsite for
recovery. Cost savings per gallon of spent Freon recovered are even higher
than in the recovery of methyl chloroform because of the extremely high cost
of virgin Freon. The total annual savings are less only because less spent
Freon is generated.

Environmental Evaluation

     The purpose of this study is to present case studies of methods of waste
management that are alternatives to land disposal. The intent was to show the
reduction in the quantity of land disposed waste achieved through use of the
alternative technology. In this particular case study, the spent solvents
have been recovered on site, in some degree or another, for over ten years.
In addition, even if they were not recovered on site, they could easily be
sent offsite for recovery. Land disposal of spent halogenated solvents of
almost 99 percent purity, particularly in the quantities that this facility
generates, is not economically practicable. Consequently, economic and not
environmental factors are the driving force behind onsite solvent recovery.

     Residues and emissions are generated by onsite distillation and
evaporation. The primary residues generated are still bottoms. These still
bottoms, contain at least 90 percent solvent, and are sent offsite for further
recovery. The quantities of still bottoms that were sent offsite during the
years 1981 through 1984 are presented in Table 33. These quantities are only
5 percent of what would be sent offsite if onsite recovery were not
practiced. One environmental benefit of onsite recovery is that a smaller
quantity of solvent is transported offsite. The chances of an accident
occuring in which spent solvent is spilled into the environment are reduced.
Another benefit is that loading and unloading the solvent from tank trucks is
greatly minimized therefore reducing the possibilities of air emissions and
spills.

TABLE 33.   QUANTITIES OF STILL BOTTOMS GENERATED FOR OFFSITE RECOVERY

Waste type	Yearly Quantity Generated			
	1981	1982	1983	1984
Methyl Chloroform				
gals.	150,000	166,000	90,000	129,000
lbs.	1,400,000	1,550,000	840,000	1,200,000
Methylene Chloride				
gals.	94,000	182,000	136,000	74,000
lbs.	1,020,000	1,970,000	1,470,000	800,000
Methyl Chloroform/Freon				
gals.	--	--	77,640	119,460
lbs.	--	--	880,600	1,355,000

Another potential environmental impact of onsite solvent recovery is emission of volatile solvents to the atmosphere. The primary source of these emissions would most likely be the vacuum pump associated with the flash evaporator. Unfortunately, it was not possible to measure these emissions. As previously shown in Figure 7, these emissions are vented to a carbon adsorption unit which has a removal efficiency ranging from 85 to 95 percent. Since the solvent recovery operations are all indoors, it is possible to vent other fugitive emissions from pumps, valves and other fixtures to carbon adsorption units. The majority of the air releases from solvent recovery are captured. After the capacity of the carbon in the adsorption units is spent, the absorbed solvents are desorbed by steam stripping, the water/solvent mixture is decanted, and the solvent fraction is reused after recovery by distillation or evaporation.

8. Facility D Case Study

Facility Description

Facility D manufactures mobile communications equipment in an operation consisting of 260 employees. Process operations consist of a small metal forming shop, prepaint and painting lines, electroplating, and electroless plating of printed circuit board components. Onsite wastewater treatment includes cyanide destruction, hexavalent chromium reduction, and acid/alkaline neutralization. Organic solvents involved in photoresist developing and stripping are recovered in-process through distillation. The waste stream of interest is a spent developing solution, consisting of 1,1,1-trichloroethane and non-stabilized Dupont Riston photoresist.

Waste Sources

Dry film photoresists such as Dupont Riston are accepted by the industry as the most reliable technology for producing printed circuit panels at high yields. The typical printed circuit panel prior to imaging consists of a one ounce per square foot of copper foil clad on a fiberglass-epoxy substrate. A complete list of Printed Circuit Board manufacturing unit operations is presented in Table 34, while the discussion below highlights those operations relevant to Facility D.

Drilling and Deburring--
After mechanical or chemical cleaning the typical double-sided or multilayered board is drilled by numerically controlled, high speed spindle drills. The resulting holes and board edges are then deburred by rotating brushes to remove any loose particulate matter or rough edges detrimental to subsequent chemical processing operations.

Electroless Copper Deposition, Rinsing and Neutralization--
The first chemical process is the deposition of a 25-85 microinch layer on the surface and in the drilled holes of the panel. This thin layer provides electrical contact from surface to surface and layer to layer. The reaction does not require electrical current and, therefore, is dependent mainly on three factors: chemical activity, mechanical agitation, and temperature. After electroless copper deposition, the board is thoroughly rinsed, neutralized with a mild acid, and dried to ensure that the surface will be receptive to the photoresist.

TABLE 34. PLANT D PRINTED CIRCUIT BOARD PROCESSING

1. Copper Clad Board

2. Board Cleaning

3. Drilling

4. Deburring

5. Electroless Copper Pretreatments

6. Electroless Copper Deposition

7. Hot Roll Lamination

8. Image Transfer

9. Developing

10. Electroplating - Copper

11. Electroplating - 60/40 Solder

12. Immersion Tin

13. Resist Stripping

14. Ammonical Etching

15. Electroplating - Tabs

16. Finished Board

Hot Roll Resist Lamination--
    Dry film photoresist is an ultraviolet light sensitive organic polymer
(methylmethacytate) applied directly to the copper foil surface by a hot roll
laminator. All surfaces must be absolutely clean because any foreign
particles laminated under the resist will cause electrical shorting if a
particle falls across a circuit. To prevent resist overhang, edges should be
trimmed flush with the panel.

Image Transfer--
    In the photo imaging operation, a stencil of the artwork is exposed to
ultraviolet light under vacuum. Vacuum maintains good contact between the
photo tool and the resist surface, preventing blurred images. All desired
features such as circuit traces, hole pads, and connector tabs remain under
the shaded area of the stencil and, therefore, unexposed. The dry film
covering the undesired areas, such as excess copper cladding, is exposed and
polymerized. Since polymerized photoresist acts as an electroplating
inhibitor, only the areas remaining unexposed will receive the required
metallic deposition in the subsequent plating operations.

Resist Developing--
    Following exposure, the printed circuit panels are introduced to a high
pressure spray of organic solvent. The solvent, 1,1,1-trichloroethane,
dissolves the unexposed photoresist and reveals the bare copper underneath.
Spent TCE containing unstabilized resist is automatically gravity fed from the
developing solution tank to a Dupont-Riston solvent recovery still.

Electrolytic Plating--
    At Facility D, the panels are racked and immersed in a MacDermid
acid-copper electroplating bath until the drilled holes, circuit traces, etc.,
acquire the sufficient plating thickness. The required 1-2 mils of copper
plating insures a uniform electrical conductivity throughout the panel.
Following copper plating, the panels are then immersed in a 60/40 Sn/Pb alloy
electroplating bath in which they receive 30-50 microinches of plate that acts
as an etch resist.

Resist Strip--
    To remove polymerized photoresist, the boards are rinsed, dried and then
run through a conveyorized stripper. The active ingredient is methylene
chloride/methanol which is applied to the board surface by spray nozzles that
penetrate into the narrow channels between circuits to remove resist.
Incomplete removal of resist will cause problems in subsequent processing,
particularly the ammoniacal etchant step.

Developer Waste Management

    The purpose of this case study is to evaluate developer solvent waste
minimization technologies, particularly the state-of-the-art application at
Facility D. The process depicted in Figure 9 represents a semi-continuous,
2-step, distillation recovery system. Until recently, all still bottoms from
the primary distillation unit, a Dupont Riston SRS-120 solvent recovery still,
were drummed and shipped offsite for reclamation. In October 1985, Facility D
purchased a Zerpa Recyclene RX-35 solvent recovery system to reclaim still bottoms onsite. A brief description of the two technologies is presented below.

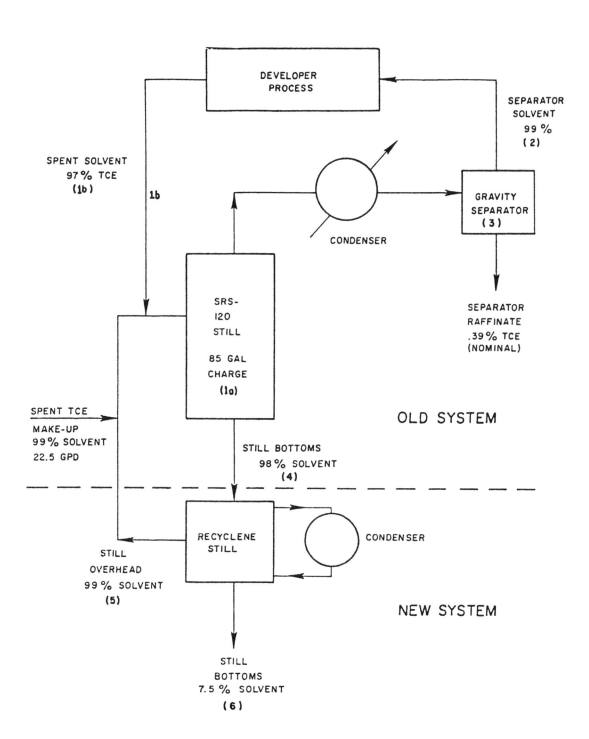

Figure 9. Plant D 2-stage solvent recovery system.

Dupont Riston SRS-120--
    The SRS-120 is a liquid phase recovery system whose operation is consistent with conventional batch distillation technology. The apparatus consists of a 100 gallon boiling chamber, an overhead condenser, and a 3 gallon water separation system. The still operates at atmospheric pressure and a boiler temperature of 165°F, which is the boiling point of 1,1,1-trichloroethane. The major recovery criteria is overhead solvent specification and contaminant content in the product. Prior to operation, the still must be filled with an 85 gallon charge consisting of overhead from the recyclene unit and any required make-up fluid. Feed from the developer solution tank is gravity fed to the still on a semicontinuous basis. The feed varies in solids content depending on the type of board being processed, but it typically ranges from 1-3 percent photoresist. The heat of vaporization is supplied by a closed coil circulating system using low pressure steam (2-15 psi and 250°F) to provide the heat input. Due to the wide range in component boiling points encountered in this type of application, this system is economically feasible for feed streams with low solids content of 5 percent or less [Pace, 1983]. The low operating temperature (165°F) insures that the photoresist will not polymerize and foul heat transfer surfaces, decreasing still efficiency.

    The second operating phase is the condensation of the solvent vapor in a simple shell and tube heat exchanger, using ethylene glycol as the cooling medium. Vapor is removed from the still as fast as it is formed without appreciable condensation or reflux. The overhead product which is virtually solids-free is gravity fed into a 3-gallon decanter to separate any water introduced in the spent feed from the product solvent. TCE from the bottom of the separator is gravity fed to a ten gallon sump where it is temporarily stored. A level control on the sump returns this product to the developing fluid holding tank which displaces a similar quantity into the developing tank, completing the closed loop system.

    Since little true fractionation occurs, increasing recovery will result in degradation of the overhead product. Therefore, after the completion of each recovery cycle (typically 2 days), the vessel is opened and the remaining sludge and solvent mixture is drained through ports in the vessel bottom. The solvent mixture, which may contain up to 8 percent solids, is then pumped into barrels awaiting transfer to the Recyclene unit.

RX-35 Recyclene Still--
    The Recyclene RX-35 solvent recovery system is a batch distillation apparatus analogous to the Dupont Riston SRS-120. The system (Figure 10) consists of a 30-gallon capacity, silicone oil immersion heated stainless steel boiler, a non-contact water-cooled condenser, and a 10 gallon temporary storage tank. The boiler is equipped with a vinyl liner inside a Teflon bag. The Teflon bag provides temperature resistance and the vinyl bag collects solid residue, eliminating boiler clean-out and minimizing sludge generation after distillation. Two thermostats control the temperature of the boiler and the vapor, automatically shutting down the boiler when all the solvent has evaporated.

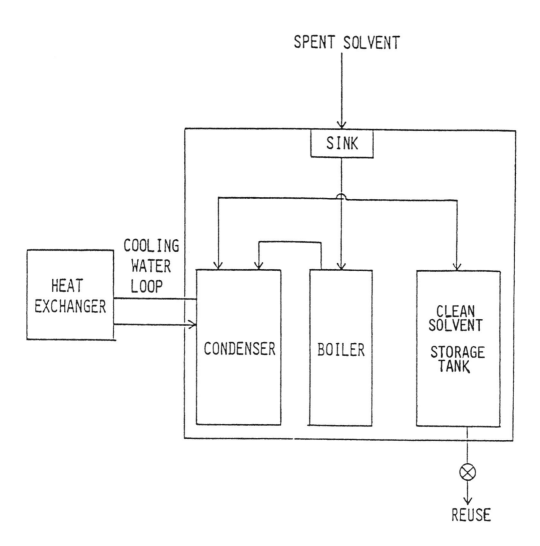

Figure 10.   Process flowsheet - Recyclene distillation solvent recovery.
Source: Dietz J.D. and Cherniak C.M., 1984.

Prior to operation, a 20-25 gallon charge is transferred into the boiler from the barrels containing the Dupont Riston SRS-120 still bottoms. The boiler which operates at 1 atm is heated to 370°F through use of an electrically heated silicone oil jacket. Cleaned solvent vapors (99.5 percent pure) rise through a water cooled condenser (85 gph) and are collected in the 10 gallon temporary storage tank. Typical batch mode operation time requires 90 minutes. The SRS-120 of Plant D is charged 4 to 6 times daily. Following complete evaporation, the bottom product is a dry solid consisting of 1,1,1-TCE, waste photoresist, and residual contaminants. The residual contaminants are believed to be trace amounts of electroless copper solution and corrosion inhibitors. Since TCE in the presence of water can liberate hydrochloric acid and react violently with aluminum, inhibitors are necessary to prevent corrosive reactions. Facility D is also recycling 42 still bottom drums which have been stored onsite in anticipation of acquiring the recyclene unit.

The truly unique feature of Recyclene RX-35 is a patented bag liner system which keeps the heat transfer surface dry and clean, consequently making it easier to operate the system (Nemec, 1984). Otherwise, contamination would result in reduced efficiency, increased energy requirements, and decreased distillation rate efficiency. The RX-35 also concentrates the waste, thereby reducing waste volume.

A few technical limitations of the Recyclene unit should be noted. The maximum operating temperature is 390°F, so that recovery of solvents with higher boiling points would not be practical. Addition of a chiller is often necessary to condense compounds with depressed boiling points.

PROCESS TESTING AND ANALYTICAL RESULTS

Process Testing

On January 23, 1986, a field study was conducted to evaluate waste minimization operations at Facility D. Sampling was conducted over the course of a normal day's operation when both the Dupont distillation and Zerpa solvent recovery units were in operation (see Table 35 for recyclene still process information). Seven separate sampling locations were utilized to provide a comprehensive process evaluation and mass balance. Those locations are: 1a) the initial Riston still charge, 1b) the spent developer solvent prior to entry into the Riston still, 2) clean solvent exiting the water separator, 3) the contents of the water separator, 4) the still bottoms product from the Riston distillation unit, 5) recovered solvent from the recyclene unit, and 6) the final bottoms product from the Recyclene unit.

Upon arrival at Facility D several discrepancies were noted between actual operation and the original process description in the QA Sampling Plan (these deviations have been corrected in the case study process description): (1) still bottoms from the Dupont Riston SRS-120 were not pumped continuously to the Recyclene RX-35 solvent recovery still, but were pumped when needed from 55 gallon storage drums, (2) Dupont Riston still was not filled continuously from the developing solution tank, but was filled with recovered TCE from storage drums, (3) no virgin TCE was used in the process. Other

TABLE 35. FACILITY D PROCESS INFORMATION
RX-35 RECYCLENE SOLVENT STILL

Parameter	Design[a]	Operation during testing
Capacity (gph)	35	25
Thruput (gph)	10-35	12
Temperature (°F)	390	370
Pressure (atm)	1	1
Overhead (%) Purity	99.5	99.9
Yield (%)	95+	99.8

[a]Source: Blodgett, W.A., 1985.

discrepancies during testing were a two hour delay in the sampling of the Dupont Riston still overhead due to a faulty valve assembly and a 15 gallon process spill.

Due to these slightly different operating conditions, allowances were made and sampling/testing proceeded in the following manner. The Recyclene still bottoms were analyzed for extractable and volatile organics, and EP Toxicity metals since it is expected that this product will be landfilled. The other six streams were sampled for volatile organic compounds, extractable organics, metals, and total solids. The spent solvent influent and Riston still overhead were sampled at 5 (instead of 8) hourly intervals over the day to provide a representative composite. The SRS-120 still bottoms was sampled at the beginning of the sampling effort while the recyclene bottoms, overhead and water separator contents were collected at the end. In addition, the principal investigator elected to sample the initial Dupont Riston SRS-120 still charge. This charge consisted of 85 gallons of Recyclene still overhead collected from previous Recyclene batch still operations. The initial Riston charge was then analyzed for volatiles, extractables, and solids to fully characterize the system. Finally, the total metals analyses proposed, were not conducted in order to reduce program analytical costs.*

During the previous day's production, 1500 (12"x18"), 2-sided boards were developed through the spray application of 1,1,1-trichloroethane. Roughly 50 percent (2,250 ft^2) of the photoresist was dissolved during this operation and then accumulated in the 85 gallon capacity of the Dupont Riston solvent recovery still. On the day of testing, the still bottoms were drained from the SRS-120 and pumped into two 55-gallon drums. The feed to the Recyclene RX-35 still consisted of 25 gallons of contaminated solvent transferred from one of these drums.

Analytical Results

The various process flow streams are primarily either solvent or dissolved photoresist (with the possibility of trace metals). Thus, the composition of the streams can be determined through both a volatile organic and total solids analysis. The results of these analyses are summarized in Tables 36 and 37, respectively. Note that a large percentage (6.7-11.0) of the total solvent mixture is composed of carbon tetrachloride. The presence of this solvent is unexplainable since it is not normally found in solvent waste streams typical to printed circuit board manufacturing. In fact, the relatively high concentrations of carbon tetrachloride in the feed and product streams came as quite a surprise to both the investigators and plant personnel. While it was determined that Facility D had not used carbon tetrachloride for some time, it is possible that it was introduced into the system from old solvents stored onsite in contaminated containers and not as a breakdown product from the recyclene still.

---

*In accordance with revised proposal to EPA Project Monitor Harry Freeman dated 28 February 1986.

TABLE 36. SUMMARY OF ANALYTICAL RESULTS FOR VOLATILE ORGANIC COMPOUNDS

Analytical results (w/w%)

Parameter	Stream 1a Riston still initial charge	Stream 1b Riston still continuous feed	Stream 2 Riston still distillate	Stream 3 Water separator discharge	Stream 4 Riston still bottoms	Stream 5 Recyclene still distillate	Stream 6 Recyclene still bottoms
1,1,1-Trichloroethane	92.0	100.0	100.0	0.39	92.0	92.0	7.5
Other solvent (total)	12.14	15.09	9.83	0.20	10.07	12.05	1.01
Methylene chloride	0.5	0.52	0.48	0.08	0.28	0.27	0.01
Acetone	--	0.49	0.17	0.01	0.19	0.22	0.01
1,1-Dichloroethene	0.64	0.64	0.78	0.01	0.11	0.66	0.98
1,2-Dichloroethane	--	0.89	0.20	0.01	0.11	0.12	0.06
2-Butanone	--	3.6	1.5	0.02	1.2	1.1	0.01
Carbon tetrachloride	11.0	8.1	6.7	0.07	8.4	9.8	0.79
Vinyl acetate	--	0.46	0.08	-	0.11	0.12	0.01
2-Hexanone	--	0.17	0.08	-	0.11	0.12	0.01
Tetrachloroethene	--	0.22	0.08	-	0.11	0.12	0.07

TABLE 37.   RESULTS OF SOLIDS ANALYSIS

Waste stream	Description	Concentration (mg/kg)
1a	Initial Riston Charge	460
1b	Riston Still Feed	1,200.0
2	Riston Still Distillate	1.7
3	Water Separator Discharge	-
4	Riston Still Bottoms	23,000.0
5	Recyclene Still Distillate	6.4

For the purpose of this case study, 1,1,1-trichloroethane and carbon tetrachloride were combined into the general category of solvent. Other components such as 1,1-dichloroethene or 2-butanone (MEK) were either 1,1,1-trichloroethane breakdown products, buffering agents, or corrosion inhibitors. The final assumption necessary for a complete mass balance is that all components less than 0.12 percent (below detection limits) are equivalent to 0. The only exceptions are 1,1-dichloroethane, 1,2-dichloroethane, and tetrachloroethane in the recyclene still bottoms.

Dupont Riston Still Characterization--
    The Dupont Riston still characterization (see Table 38) consisted of sampling the initial Riston still charge (Stream la), the Riston still feed (Stream lb), the distillate (Stream 2), and the water separator discharge (Stream 3). The initial Riston charge (IRC) which was described previously, was analyzed to contain 98.81 percent solvent, 1.14 percent "other" volatile components, and 0.05 percent total solids. The solids concentration in the IRC was greater than expected, due to a resaturation of the solvent by polymerized photoresist which had collected on the sides of the still. Solids were continually added during the distillation process by the feed stream which contained approximately 99 percent volatiles and 0.12 total solids. Distillation of the IRC and feed Stream resulted in a clear overhead which contained 96.62 percent solvent, 3.37 percent other volatiles, and less than 0.0002 percent solids. The water separator discharge which was sampled at the end of the day was found to contain virtually no solids, 0.46 percent solvent, and the remainder was other components, primarily water.

TABLE 38.   DUPONT RISTON SRS-120 SYSTEM CHARACTERIZATION

	Concentration (wt %)			
Parameter	IRC (Stream la)	Influent (Stream lb)	Distillate (Stream 2)	Water separator discharge (Stream 3)
Solvent	98.81	92.89	96.62	0.46
Solid	0.05	0.12	0.0002	-
Other	1.14	6.99	3.37	99.54[a]

[a]Consists of 99+ percent water

Recyclene Still Characterization--
    While the sampling and analytical assessment dealt with the entire two stage solvent distillation system described earlier, the primary focus of this case study is the performance of the Recyclene batch still. Thus, streams of primary interest include Stream 4 (Riston Bottoms/Recyclene Feed), Stream 5 (Recyclene Distillate) and Stream 6 (Recyclene Bottoms). Table 39 details the loading distributions calculated from the solvent mass balance for the

parameters of interest. The distillation of the contaminated solvent resulted in a clear overhead containing 92.8 percent solvent, 7.1 percent other volatile components, and less than 0.001 percent of total solids. The low accumulation of non-volatiles (solids) in the distillate resulted in a total overhead purity of 99.99[+] percent and a volatile component yield of 99.78 percent. In comparison, manufacturer's specifications for the Recyclene RX-35 solvent still were 99.5[+] percent and 95[+] percent, respectively. Approximately 2.5 percent of the initial solvent charge was recovered as bottoms product, with only 8.3 percent (0.60 lb) of the residual weight classified as solvents. This represents a 97.5 percent decrease in waste volume generation and a significant (99.8 percent) reduction in non-fugitive emission related solvent losses.

TABLE 39.   RECYCLENE STILL MASS BALANCE

Parameter	Loading (lb/batch)[a]		
	Influent	Distillate	Bottoms
1,1,1-trichloroethane solvent	255.20	254.60	0.60
Solid	6.47	0.01	6.46
Other volatiles	19.63	19.61	0.02

[a]Based on 25-gallon charge.

Process Residuals--
     Since it is expected that the residual bottoms product will be disposed of through land disposal, the manufacturer's claim was investigated, that in some cases the RX-35 will convert hazardous residue to nonhazardous residue. (Nemec, Nov./Dec., 1984) This goal was accomplished through an EP Toxicity Metals analysis, an organic extractables analysis and a volatile organic analysis. Table 40, which compares Plant D's EP Toxicity Metals results to Federal guidelines, shows that the bottoms product is well within current compliance standards and low on metallic contaminants. Organic extractables results for Plant D were all below detection limits (see Table 41).These low concentrations (less than 0.0003 weight percent) are too minute to be included in the process flow stream characterization. Therefore, the bottoms product (with the exception of 1,1,1-trichloroethane) is assumed to contain no or very little priority pollutants. The volatile organic analysis, previously presented in Table 36, show the recyclene bottoms product (Stream 6) contains 7.5 percent by weight of 1,1,1-trichloroethane. This concentration classifies the recovery process residue as a F002 (trichloroethane recovery still bottom) toxic hazardous waste.

TABLE 40.   PLANT D EP TOXICITY METALS RESULTS FOR
           RECYCLENE STILL BOTTOMS

Element	Concentration (mg/L)	EPA standards (mg/L)[a]
Arsenic	0.03	5.0
Barium	0.106	100.0
Cadmium	0.002	1.0
Chromium	0.003	5.0
Lead	0.03	5.0
Mercury	0.0009	0.2
Selenium	0.04	1.0
Silver	0.01	5.0

[a]U.S. EPA, Federal Register, V.45, No. 98:33122,
May 14, 1980.

TABLE 41.   FACILITY D SEMI-VOLATILE ANALYSIS RESULTS

Sample I.D.	Description	Results (mg/kg)
Stream 1a	Initial Riston Charge	< 0.76
Stream 1b	Riston Still Feed	< 0.75
Stream 2	Riston Still Distillate	< 0.75
Stream 3	Water Separator Discharge	< 0.20
Stream 4	Riston Still Bottoms	< 15
Stream 5	Recyclene Still Distillate	< 0.76
Stream 6	Recyclene Still Bottoms	< 500.0

ECONOMIC AND ENVIRONMENTAL EVALUATION

Economic Evaluations

One of the goals of this program was to evaluate the economic practicability of the Recyclene RX-35 batch still at Plant D. Based on the recyclene still mass balance (Table 39), the quantity of waste generated prior to the installation of the RX-35 unit was roughly 10,625 gallons per year. Operation of the 2-stage solvent recovery system resulted in the recycling of 10,602 gallons per year of solvent not lost through fugitive emissions. This figure represents a 97.5 percent reduction in waste volume and a 99.8 percent recovery of solvent in the overhead. While these results are very encouraging, the economic feasibility of the recyclene process will ultimately determine the extent to which it is applied.

The capital costs for the RX-35 which total $26,150 include the installed purchase price for the basic unit ($25,850.00) and a start-up service fee ($300.00). Differential energy consumption was calculated on the basis of 425 batches, 47 Kwh per batch at $0.06 per Kwh. In addition to electricity, operating expenses at Plant D include labor (one manhour per batch) at a cost of $3,120 and liner consumption (2.7 batches per nylon liner) for a total cost of $3,358 per year. Finally, 3.2 tons of residual solids were estimated to be disposed of through landfilling, at a cost of $200.00 per ton. Thus, the total first year cost for implementation and operation of a Model RX-35 with auto-fill at Plant D was found to be $34,473.

Table 42 lists the annual cost savings and waste reduction calculated for Plant D. Approximately 10,600 gallons of solvent were estimated to be recycled in the first year of operation. This represents a disposal savings, at $0.35 per gallon of solvent, of $3,710 per year. However, more substantial is a savings of $47,709 per year in virgin solvent purchases (at $4.50 per gallon). When the two savings totals are summed, the aggregate annual savings is $51,428. This figure represents a net first year savings of $16,955 (includes total capital cost) and an estimated investment payback period, after considering credit for reclaimed solvent and reductions in waste transportation and disposal costs, of 7.3 months.

Environmental Evaluation

While the Recyclene RX-35 solvent recovery still does significantly reduce the volume of hazardous waste generated, the RX-35 does not eliminate hazardous waste completely. In the advent of a total ban on the land disposal of hazardous solvent wastes, alternate methods of disposal, such as solidification or incineration, would have to be investigated.

TABLE 42.   ANNUAL COST SAVINGS AND PAYBACK FOR RECYCLENE RX-35 AT PLANT D

Cost Item	Number of Units (per yr)	Cost per Unit ($)	Cost Prior to installation ($)	Cost after Installation ($)
Contaminated Solvent	10,625 gal	0.35	3,719	-
Recyclene Bottoms	3.2 tons	200	-	640
Differential Solvent Purchase	10,602 gal	4.50	47,709	-
Differential Energy Consumption	20,092 kwh	0.06	-	1,205
Replacement Liners Teflon	52 bags	45.15	-	2,348
Nylon	155 bags	6.50	-	1,010
Additional Labor	208 hrs	15.00	-	3,120
TOTAL COST			51,428	8,323

ANNUAL COST SAVINGS (1st year)   43,105
RECYCLENE RX-35 PURCHASE AND INSTALLATION COST   26,150
PAYBACK PERIOD   7.3 mo.

9. Facility E Case Study

Facility Description

Facility E began operations in January 1982 as a manufacturer of customized, fine-line multilayer printed circuit boards. Facility E utilizes a subtractive process to produce boards with up to 22 layers, which are then shipped to other facilities for assembly operations. The plant employs 600 people and has an annual production volume of 600,000 ft^2 of finished boards. Production volume is expected to double within the next few years when the facility completes planned construction of an additional plating line. Facility E currently operates 5 or 6 days/week, 24 hours/day. Its onsite wastewater treatment plant operates 7 days/week.

Facility E initiated an ambitious waste minimization program in mid-1984. Since that time, production has roughly doubled, but liquid discharge to the wastewater treatment plant has remained constant and wastewater sludge generation has dropped roughly 30 percent. Waste minimization efforts continue to center around in-process modifications to use nonhazardous or reclaimable solutions, to reduce water consumption and bath dump frequency, and to optimize wastewater treatment operations. These programs and a description of Facility E's production and waste treatment processes are described below.

Facility E treats all process rinse waters and spills in the onsite wastewater treatment plant. Aqueous process baths are treated in-line, in the treatment plant, or are temporarily stored in a tank farm to be reclaimed offsite. The only solvent used in significant quantity in the plant is 1,1,1-trichloroethane (TCE), which is used as a presolder mask cleaning agent. This is recovered onsite in a still equipped with provisions for secondary recovery of solvent from drummed still bottoms. Solid hazardous wastes include wastewater sludge, TCE still bottoms, spent activated carbon solids, and potassium hydroxide resist stripper sludge. Other filtered solids, filter paper, and waste board materials (15 to 25 percent of production) are nonhazardous and, therefore, disposed in a sanitary landfill. Through the use of nonhazardous and reclaimable process solutions, Facility E has significantly reduced the quantity of waste which would otherwise have to be disposed of as hazardous material. A summary of the primary waste streams is provided in Table 43. More detailed description of their origin, disposition, and constituents of concern is provided below under the process area in which they are generated.

109

TABLE 43.   FACILITY E PROCESS WASTE SUMMARY

Chemical name	Use in process	Annual quantity gal/yr (lb/yr)	Disposition	Comments
Ammonium Chloride	Etchant	208,000	Reclaimed and supplied by CP Chemical in Sumter.	Hazardous waste. Replaced CuCl. Facility E plants to use Mercer process (2-stage solvent extraction, electroventing Cu) for acid reclamation in the future. Contains 11% Cu with NH_4OH, OH, NH_4Cl, and $(NH_4)_2CO_3$.
Nitric Acid	Rack stripping in electroless and electroplate operations.	8,100	Reclaimed by SCA (GSX) in NJ.	Will treat onsite beginning in January. Contains Pb from solder plating.
Sulfuric Acid	Epoxy smear removal in electroless plating area.	24,000	Reclaimed by City Services.	Exemption as recyclable material. 93% pure in spent solution (98% in virgin). Planning to discontinue use in future.
Metal Hydroxyde Sludge (F006)	Wastewater treatment plant sludge.	(1,040,000)	Landfilled as a hazardous waste.	Contains Cu, Ca, Na, and acrylic photoresist. Small amounts of Pb, Sn, Ag, Mn and high pH. No Cn.
Potassium Cyanide Solution	Gold stripping (closed process).	---	Recovered offsite.	
1,1,1-trichloroethane Still Bottoms	Solder mask cleaning agent.	220	Landfilled as a hazardous waste.	TCE is recovered onsite in a closed loop, 2-stage distillation process (90% recovery).
Potassium Hydroxide Sludge/Filter Paper	Inner layer resist stripping solution sludge collected on paper filters.	8,800	Landfilled.	High pH. Consists primarily of filter paper (95%) and resist (acrylic polymer). Solidified with lime.
Potassium Carbonate Developing Solution	Inner layer and outer layer.	---	Wastewater treatment.	Resist is too hydrolyzable to settle by gravity.
Scrap Boards, Edge Cuttings	In all process areas.	(129,000 s.f.)	Sanitary landfill.	Cu, Sn, Pb on boards. Looking for reclaimer.
Pb/Sn Solder Plating Bath	Pattern plating.	220	Recovered.	Send to vendor as hazardous waste and reclaimed.

(continued)

TABLE 43 (continued)

Chemical name	Use in process	Annual quantity gal/yr (lb/yr)	Disposition	Comments
KOH Resist Stripping Solution Sludge	Outer layer resist stripper.	4,300	Landfilled as a hazardous waste.	Resist is settled by gravity and consolidated in an inclined hydro-sieve filter. The sludge is dewatered to 15% solids in a holding bag prior to drumming. Hazardous due to Cu (10% $CuSO_4$), Pb and pH.
Fluorinert/Filters	Fuse pre-clean.	---	Recovered offsite.	Recovered by supplier.
Sodium Hydroxide	Fuse pre-clean.	12,870	Wastewater treatment.	
Filtered Solids and Filter Paper	Primarily in electroless and pattern plate.	2,200	Landfill.	Filtered solids are neutralized prior to landfilling.
Hydrochloric Acid/ Thiourea	Fuse pre-clean.	15,730	Wastewater treatment.	
Na_2SO_4 Micro-Etch	Electroless and micro-plate.	57,860	Wastewater treatment.	Electroless etch 9200 gal tank) is dumped once per day. Cu concentration ranges from 10,000 to 50,000 ppm. Micro-plate etch is dumped in small quantities; 1 drum every month.
Monoethanolamine Stripping Solution	Pattern plating and outer layer strip.	16,730	Waste treatment.	Used for in-line stripping to correct image errors.
Potassium Carbonate Developer	Solder mask.	Variable.	Waste treatment.	

Waste Sources

The production facility houses administration offices, raw material
storage, a wastewater treatment operation with a wet/dry laboratory, and
production lines. Shipping/receiving and a tank farm (21 tanks) are located
on the northeastern end of the production building. The onsite wastewater
treatment plant processes 280,000 to 300,000 gpd of complexed and noncomplexed
rinsewater and process baths in both flowthrough and batch systems. The
manufacturing facility includes separate processing areas for inner and
outer-layer operations. The process is described below with reagent usage and
waste generation discussed for each processing area.

Board Cutting/Inspection--
Facility E uses two-sided, copper foil clad, epoxy/glass cloth boards as
the base material for its printed circuit boards. Reject boards are disposed
in a sanitary landfill, along with any unreclaimable boards which are
defective due to improper processing. Facility E does not currently have
plans to investigate recovery options for these boards.

Inner Layer Chemical Clean--
Boards are chemically cleaned in two processing lines. The first line
employs sulfuric acid (H_2SO_4) and Metex E-250 (50 percent potassium
hydroxide; KOH), and the second uses hydrogen peroxide/sulfuric acid baths
with copper sulfate ($CuSO_4$) and a stabilizer (sodium salts and phosphoric
acid). These solutions and rinsewaters are sent to the waste treatment plant.

Inner Layer Image--
Boards are spray cleaned with 10 percent H_2SO_4, mechanically
scrubbed, and air dried. Resist (Dynachem Film Laminar TR, containing
methylmethacrylate) is applied in a dry film laminator using rollers and then
exposed to UV light. The cleaning solution is discharged to the waste
treatment plant.

Inner Layer Develop, Etch, and Strip--
Resist is developed by dipping in three tanks containing 1 percent
potassium carbonate (K_2CO_3). This solution is discharged continuously to
waste treatment. Other process baths are filtered in-line using spiral-wound,
Sethco particulate filters to minimize bath dump frequency. The nonhazardous
solids and filter material are neutralized and disposed in a sanitary
landfill. Together with filtered solids from other processing areas,
2,200 gallons of this waste are disposed annually. All liquid discharges to
waste treatment result from displacement of used solution in the tanks by
addition of make-up, which is added automatically to maintain necessary bath
characteristics (e.g., pH). This general arrangement is used in most other
discharged process baths. Counterflow rinse water is used to generate the
developer make-up solution. This conservation effort alone has resulted in a
20 percent reduction (5 gpm) in water requirement in this area.

Ammonium chloride (NH_4Cl) is currently being used as the board
etchant. This solution contains 12 oz. of Cu/gallon, as well as ammonium
hydroxide and ammonium carbonate. Etchant is used at a rate of roughly 2 gpm
and is reclaimed offsite by MacDermid, the raw material supplier. Together

with spent outer-layer etchant, over 2 million pounds of this waste are generated annually. Facility E is currently exploring options for onsite recovery (Mercer Process).

A 2.5 percent solution of potassium hydroxide (KOH) is used as the inner layer stripper. Spent solution is continuously fed to two in-line gravity paper filters equipped with automatic paper advance. Approximately 160 drums of solid waste is generated annually consisting of 95 percent filter paper, and only 5 percent acrylic polymer sludge. It is neutralized and solidified with lime prior to landfilling as a hazardous waste. At this juncture, Facility E has not identified an alternative technology to effect separation of the highly hydrolizable resist. They are investigating the use of fine-mesh, reusable filters.

Inner Layer Surface Treatment--
Inner layer boards undergo surface preparation prior to lamination. A bronze oxide/potassium hypochloride solution is used to generate a rough copper oxide layer which prevents peeling when the board is laminated. Cleaning solutions contain KOH, H_2SO_4, NaOH, and $NaClO_2$. Since a small quantity of copper is stripped off during surface treatment, the line is equipped with a counterflow recovery rinse. Boards are air dried prior to lamination. No chemical drying agents are used in this facility.

Lamination, Drill, and Deburr--
The only waste generated in lamination consists of fines (epoxy, acrylic, some copper) which are filtered out of a recirculating water wash which keeps the laminator clean. These fines are generated in small quantities and dumped in a sanitary landfill.

Electroless Copper Plating--
Acid/alkali solutions are used for cleaning, rinsing, conditioning, and activating the board surface for palladium catalyst deposition. These solutions contain Na_2CO_3, H_2SO_4, NaOH, NaF, $KMnO_4$ (residue oxidizer), HCl (activator), $SnCl_2$, hydrazine/H_2SO_4 (accelerator), $PdCl_2$, and organic activators such as ethanolamine. Many of these solutions are proprietary mixtures supplied by MacDermid. The residue oxidizer ($KMnO_4$) is dumped to a waste treatment complexed solution system in small batches (180 gallons) where it is used to oxidize complexes.

A Na_2SO_4 micro-etch is used to prepare the surface for catalyst application. This 200 gallon tank has copper concentrations of 1 to 5 percent and is dumped once each day to the wastewater treatment plant. Facility E is currently experimenting with electrolytic recovery of copper from this bath, but has not yet identified a viable method.

Activator and accelerator solutions used for catalyst application are reclaimed offsite by the vendor. Citric acid and H_2SO_4 washes, which precede the electroless copper plating tanks, are discharged to wastewater treatment. Electroless baths are proprietary solutions containing copper salts, formaldehyde, small quantities of CN (4 ppm), organic chelators (e.g., EDTA), and NaOH. These electroless baths are reclaimed offsite by MacDermid. Countercurrent rinses and in-process filtration are used to reduce water consumption and extend bath life, respectively.

Outer Layer Image Transfer--
   Boards are spray cleaned with recirculated 10 percent H_2SO_4, mechanically scrubbed, and air dried prior to image transfer. Dupont Riston 3620 (contains methacrylates) is applied in a dry film laminator and developed through exposure to UV radiation. Sulfuric acid solutions are discharged to wastewater treatment.

Outer Layer Developing--
   Potassium carbonate (K_2CO_3) in a 6.5 percent solution is used as the outer layer developing agent. This solution is continuously discharged to waste treatment as it is displaced by make-up fluid which is added to maintain pH. This line is equipped with a KOH strip tank which is used to reclaim boards with image errors. The outer layer developing process is currently being upgraded by Facility E to minimize water consumption in similar fashion to its inner layer counterpart (e.g., reuse rinse water in developing fluid make-up).

Pattern Plating--
   Boards are pattern plated with eight acid copper and one aqueous lead/tin plating baths in a 48 tank plating line. The line begins with a nitric acid (HNO_3) rack strip tank. Spent acid is hazardous due to its lead content from solder plating. This is combined with HNO_3 rack strip from the electroless line and reclaimed (8110 gpy) offsite. These tanks are filtered continuously to reduce dumping frequency (twice per year). Filtered solids are neutralized and disposed.

   After the racks are stripped, boards are loaded and then undergo rinsing, cleaning with phosphate solutions (H_3PO_4, Electroclean PC2000), and more rinsing before being plated. Acid copper baths contain $CuSO_4$, sulfuric acid, an organic brightener, and chlorides with copper concentrations of 12 oz/gallon. The solder plating bath contains HBO_3, BF_3, $Pb(BF_4)_2$, $Sn(BF_4)_2$, and organic acids. The general processing procedure is to activate the board surface (HCl), plate, clean/rinse, and replate.

   Cleaning baths are continuously filtered and discharged to wastewater treatment by make-up fluid displacement. They are sent out for reclamation when copper levels reach 1 to 1-1/2 lb/gallon as determined by in-process monitoring. All rinses are countercurrent flow and are discharged to waste treatment.

   Copper and solder electroplating baths are treated with activated carbon once every 3 months and every month, respectively. These electroplating baths never have to be dumped with this arrangement under normal processing conditions. The activated carbon treatment process is described in detail in the next section.

Outer Layer Strip and Etch--
   Ammonium chloride (NH_4Cl) is also used as the outer layer etchant and is reclaimed in similar fashion to the inner layer etchant. Hydrochloric acid in a 10 percent solution is used as a post-etch solder activator and cleaning solution and is discharged to wastewater treatment. Potassium hydroxide (KOH) in a 5 to 20 percent solution is used as the resist stripper. It is

continuously recovered through gravity separation of the resist from the
solution in a Hydro-Sieve inclined cascade filter without requiring any
chemical addition. Filter sludge is collected, dewatered by gravity to
15 percent solids, solidified with lime, and disposed at a rate of
4,300 gallons per year in a hazardous waste landfill. The resist sludge is
high in Cu (10 percent $CuSO_4$), Pb, and pH. Potassium hydroxide (KOH)
solution is discharged to waste treatment as it is displaced by make-up fluid.

Fuse-Preclean--
    Solder is fused in a Fluorinert vapor blanket in a completely closed
system. Fluorinert is a proprietary, long aliphatic carbon chain containing
fluorine. The system is equipped with filters which are reclaimed by the
chemical supplier. The vapor blanket is followed by a spray cleaner
containing 10 percent NaOH and a finishing solution spray containing
10 percent HCl with thiourea. These spent solutions are discharged to
wastewater treatment at a rate of 45 gpd and 55 gpd, respectively.

Microplate--
    Facility E plates nickel and gold on board tabs in a microplate line.
Tabs are micro-etched with a sodium persulphate solution (contains $CuSO_4$ and
H_2SO_4), which is discharged to wastewater treatment in small quantities
(one drum each month). Tabs are then nickel-plated using a $NiSO_4$ bath.
Sulfuric acid (H_2SO_4) and $NiCO_3$ are added for pH adjustment, boric acid
(HBO_3) is added as a buffer for nickel salts, and organic sulfates/aldehydes
are added as a stress reducer/brightener. Gold is plated in a gold cyanide
bath containing KCN and citric acid. Gold plating and rinse solutions are
recovered in an adjacent line. The recovery process uses a stripping solution
containing Technistrip Au II (aromatic hydrocarbons) and KCN and ionic
exchange equipment. Recovered rinse water is returned to the process and
spent ionic resin is reclaimed offsite. In addition, a gold recovery line has
also been installed adjacent to the wastewater treatment plant for stripping
gold off tabs of reject boards.

Solder Mask--
    Solder masking is accomplished by first cleaning with
1,1,1-trichloroethane (TCE), followed by dry film application and aqueous
solution developing. TCE is recovered in a closed-loop still, which operates
with a recirculation flow of 1 gpm for a 6 to 8 hr/day to yield greater than
95 percent recovery. The system includes solvent storage tanks for used
product and for virgin make-up which deliver solvent into the still and
process tank, respectively. The still operates automatically when it receives
a charge from the spent TCE tank. Heat is supplied to the still through a
Teflon heat exchanger and noncontact cooling water is used in the Teflon
condenser. Still bottoms are dumped roughly once each month into a drum
enclosed in a heating jacket for second-stage recovery of solvent. Although
the liquid content of the waste could be reduced further, Facility E fills and
disposes one drum of these bottoms every 90 days in order to comply with
hazardous waste drum storage regulations. Approximately one drum of make-up
TCE is added to the system each month.

Electroplating Bath Waste Management

Fundamental to the success of any modern printed circuit board is the certainty that electroplated deposits will withstand the forces and stresses that the board will encounter. For example, thermal changes during soldering and power-up/power-down place stresses on the electroplated layers which can cause cracks or failures. These failures are often the result of organic contamination from addition agent breakdown products. Multilayer boards of the type manufactured by Facility E are regulated by MIL-STD-55110 which prohibits such failures in the finished product. To prevent the loss of military certification, printed circuit board manufacturers lacking a bath regeneration system, would typically be forced to either discharge the spent plating bath in wastewater treatment, or send it offsite for reclamation.

The purpose of this case study is to evaluate the extension of electroplating bath lifetimes (and subsequent waste reduction) by activated carbon removal of organic brightener breakdown products. The acid copper baths were selected for study since recovery of this solution results in the most significant amount of waste minimization.

Prior to the discussion of the regeneration technology, it is useful to discuss the composition, function, and limitations of organic brightener systems. Organic addition agents are a blend of leveling, carrying, and ductilizing compounds. The carrier component, a high molecular weight carbon-oxygen compound, acts as a plating inhibitor to prevent overplating and burning in high current density areas. The leveling agents are often amine compounds or heterocyclic sulfur compounds required to eliminate small-hole wall imperfections due to drilling. The brightening (grain refining) compounds are usually complex reaction products of sulfur and nitrogen compounds used to improve the overall appearance of the deposit, tensile strength, and ductility. Good tensile strength (usually greater than 40,000 psi) and elongation (greater than 10 percent) are required for the deposit to withstand thermal cycling and thermal stress tests.

Organic addition agents, which ensure optimal physical deposit properties, must be stable under conditions of high agitation, current density, and solution temperatures. Unstable brightener additives frequently break down and may become incorporated into the copper deposit adversely affecting the deposit properties. Large volumes of production work on a continual basis will significantly shorten the lifespan of the addition agents and, consequently, the plating solutions. Therefore, at Facility E, it is necessary to treat the electrolytic copper baths with activated carbon approximately every 3 months.

A secondary source of organic contamination in acid copper baths is the breakdown of photoresist during the pattern plating operation. As previously mentioned, photoresists are light sensitive, organic thermoplastic polymers which harden (polymerize) upon exposure to ultraviolet light. Incomplete exposure, developing, or rinsing can result in a defective, nonstable resist which can then break down and dissolve into the plating bath. While this problem is less prevalent than organic brightener breakdown, it is still detrimental to overall plating quality. This residue may also be removed through activated carbon treatment.

At Facility E, activated carbon treatment is performed in a batch mode for acid copper, solder, and nickel microplating baths in three separate systems. These systems consist of a holding tank, mixing tank, and MEFIAG paper-assisted filtration unit. For acid copper treatment, 2,400 gallons of contaminated solution is pumped into a 3,000 gallon mixing tank. Hydrogen peroxide is added to oxidize volatile organic species and the temperature of the bath is maintained at 120 to 130°F for 1 hour. Powdered activated carbon (80 pounds) is added and the contents are mixed for 3 to 4 hours, allowing sufficient time for adsorption of the organic breakdown products.

The solution is prefiltered by diatomaceous earth, followed by recirculation through a paper-lined MEFIAG filter to remove the suspended activated carbon. The filter solids and paper are removed as needed when a predetermined pressure drop across the filter is reached. When the bulk of the activated carbon has been removed (generally after three passes of the solution through the filter), the filter is precoated with approximately 0.67 cu ft (5 gal) of diatomaceous earth. The partially treated solution is further recirculated through the filter until a particulate test indicates sufficient solids removal (no residue detected on visual examination of laboratory filter paper). Total spent solids from plating bath purification is approximately 3.5 cu ft per batch (about 1-1/2 drums every 3 months) which is disposed in a sanitary landfill. A schematic of the activated carbon treatment process is presented in Figure 11.

PROCESS TESTING AND ANALYTICAL RESULTS

Process Testing

The plating bath carbon reclamation technology at Facility E was tested during the week of February 10, 1986. Testing of the electrolytic recovery system described in the QA Project Plan was not conducted for two reasons: 1) Facility E had inadvertently dumped the static rinse batch which was critical to the sampling; and 2) delays at Facility E would make testing of the new system difficult under the time frame for this program. Sampling activities conducted at the site were limited to the bath reclamation system as discussed below.

Samples for the activated carbon treatment system evaluation were taken from the 3,000 gallon agitated treatment tank and the 30 gallon MEFIAG activated carbon filter. A brief discussion of how each waste stream was sampled is presented below:

- Contaminated Copper Solution - Samples of the contaminated copper solution were collected from the top of the continuously stirred treatment tank at approximately 11:00 p.m. The contaminated copper solution sample, obtained from the tank by use of a plastic ladle, was taken by one of the plant employees. Duplicate samples of volatile organics, extractable organics, metals and TOC , were taken from the tank at this time. These samples were held in a polyethylene sample cooler until the sampling activities were completed the next day.

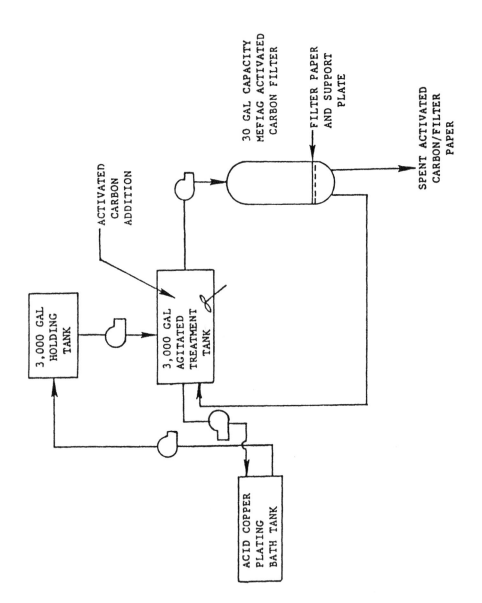

Figure 11. Facility E activated carbon treatment system.

●   Spent Activated Carbon - Spent carbon samples from the filter press were field composited for metals, extractable organics, and TOC over the duration of the treatment operation. Volatile organic samples were taken at discrete sample times. The composite samples were volume composited, based on the approximate percent recovery of carbon found in each filter run. The first sample composite contained diatomaceous earth which is used as a prefilter. Carbon was scraped off the filter paper and into the bottle. Samples were taken for volatile organics, EP TOX, and metals. Two samples of diatomaceous earth were also taken for metal analyses.

●   Clean Copper Solution - The treated-filtered copper solution was taken from a valve on the feed line to the clean storage tank. Samples for extractable organics, metals, TOC and volatile organics were collected at the end of the 18-hour run.

●   At the completion of sampling, all samples were placed in coolers with ice and vermiculite and shipped to the analytical laboratory by Federal Express.

A complete summary of the measurements, parameters, and observations recorded during the sampling period are listed in Table 44. In addition to the sampling matrix already outlined in the Quality Assurance Project Plan, a cyclic voltaic stripping (CVS) analysis was performed by the SEL-REX division of the OMI Corporation. Since SEL-REX is the manufacturer of the 70/30 acid copper bath used by Facility E, they routinely conduct this analysis as a customer service to determine brightner concentration.

Analytical Results

As previously stated, the test plan was designed to characterize the plating solution and determine the effectiveness of activated carbon for plating bath regeneration. The sampling parameters examined were trace metals, total organic carbon, semi-volatiles, volatiles, and CVS. A discussion of the results for each analytical parameter is presented below.

Trace Metals--
The objective of the trace metals analysis was to determine, using atomic absorption analysis, the concentration of metallic ions at each sampling location. The resultant data allows both a mass balance and spent activated carbon characterization to be determined. Table 45 presents a raw data summary and describes the metallic ion loading for the contaminated and filtered plating solution, as well as the activated carbon residue. The initial copper loading of the contaminated plating solution was 471 lbs or 12 oz/gallon of copper sulfate ($CuSO_4 \cdot 5H_2O$). The filtered/carbon treated plating solution contained 461.9 lbs of copper or approximately 11.8 oz/gallon of copper sulfate. This represents a loss of copper to the spent activated carbon of only 9.4 lbs or 2 percent of the initial metallic copper charge. Since the recommended range for electrolytic grade acid copper sulfate solutions operating in the 0 to 40 amp/ft^2 range is 9.3 to 13.4 oz/gallon, the metallic copper loss was not detrimental to plating specifications.

TABLE 44.   SUMMARY OF PROCESS OBSERVATIONS AND SAMPLING AT FACILITY E

Time	Process observation measurement
11:50 p.m.	Sampled spent acid copper plating bath.
12:00 a.m.	Began heating bath via steam.
00:52 a.m.	Add 4 gallon 50% (H_2O_2).
01:00 a.m.	Reached process temperature of 138°F.
03:30 a.m.	Add 88 lbs powdered activated carbon.
08:00 a.m.	Load diatomacious earth onto filters from holding tank sludge.
08:05 a.m.	Pause for electrical repair.
08:25 a.m.	Filtration begins with pressure at 20 psig.
08:35 a.m.	Checked return which was gray/black in appearance.
09:30 a.m.	Checked pressure - 20 psig.
09:35 a.m.	Break down filter, drainage emptied to holding tank.
09:40 a.m.	Sampled first filter run. Composited 50% of sample volume from top filler paper. Sampled VOAs.
10:08 a.m.	Start second filter run.
10:12 a.m.	Checked pressure - 20 psig.
11:30 a.m.	Break down filter.
11:35 a.m.	Sampled second filter run. Composited 30% of sample volume from top filter paper. Sampled VOAs.
12:00 noon	Began third filter run.
01:15 p.m.	Break down filter.

(continued)

TABLE 44 (continued)

Time	Process observation measurement
01:20 p.m.	Sampled third filter run. Completed sample composed from top two filter papers. Sampled VOAs from third and fourth filter papers.
01:45 p.m.	Began fourth filter run.
02:45 p.m.	Checked pressure - 20 psig.
02:50 p.m.	Sampled virgin diatomaceous earth.
03:25 p.m.	Break down filter. Unable to sample since all but carbon fines had been removed.
03:50 p.m.	Slurry diatomaceous earth.
04:00 p.m.	Coat filters with diatomaceous earth for final run.
05:50 p.m.	Sampled clean bath. Completed testing.

TABLE 45. FACILITY E TRACE METAL SUMMARY AND MASS BALANCE

| Parameter description | Sample ID and description | | | |
	ACT-1 Contaminated solution	ACT-2 Spent activated carbon	ACT-3DIA Virgin diatomaceous earth	ACT-3M Filtered solution
Concentration (mg/L)[a]				
Copper	21,500	107,000	95.6	21,400
Lead	1.1	99	57	0.66
Tin	6.9	420	710	8.3
Loading (lbs/batch)				
Copper	471.33	9.42	–	461.91
Lead	0.024	0.009	–	0.015
Tin	0.151	0.037	–	0.114

[a]ACT-2 and ACT-3DIA concentrations are µg/g.

In addition to metallic copper, the contaminated plating solution contained small amounts of other trace metals such as tin and lead. These metals are not recommended for optimum plating performance and represent a source of inorganic impurities. Foreign anions such as tin and lead are incrementally introduced into a plating solution throughout the solution's operational lifetime. Common sources of metallic ion contamination include leaching of parts, tanks and racks, or drag-in from previous plating operations. If allowed to accumulate, these inorganic impurities will detrimentally effect plating quality in the following manner:

- increased resistance to flow of current;

- decreased bright range;

- increased tendency to burn;

- rough and pitted deposits; and

- reduced covering power.

While the primary purpose of the activated carbon filtration process at Facility E is to remove organic molecules, the spent activated carbon data in Table 45 shows that inorganic impurities are adsorbed as well. This co-adsorption of inorganic contaminants has the net effect of reducing total lead and tin loadings in the filtered solution (37.5 and 24.5 percent, respectively), as well as extending bath life and improving plating performance. Therefore, it can be concluded that while activated carbon treatment does remove a small quantity of divalent copper (approximately 2 percent of the bath content), the co-adsorption of inorganic impurities such as tin and lead, is beneficial.

The final objective of the trace metals analysis was to determine the suitability of the spent activated carbon residue as a landfilled hazardous waste. Spent activated carbon, which is currently classified as a hazardous waste, is generated at the rate of 0.67 cu ft per batch. However, the results of the EP Toxicity leachate test shown in Table 46 demonstrate that the spent activated carbon contains metals concentrations which are within Federal guidelines. This low toxic metals content, combined with the relatively nonhazardous nature of the organics (thiocarbamoyl-thio-alkane sulfonate class), many render the residue suitable for delisting.

Total Organic Carbon--
The objective of the Total Organic Carbon (TOC) analysis was to determine the overall organic carbon removal efficiency by the periodic oxidation and activated carbon filtration system at Facility E. However, as previously stated, brightener compounds are in the class of thio-carbamoyl-thio-alkane sulfonates and are complex reaction products of sulfur groups (thiols) and nitrogen compounds (amines). As such, these reaction products are usually oxidized by the addition of hydrogen peroxide and volatilized during the subsequent elevation of bath temperature from ambient to 120 to 130°F. The apparently low removal efficiencies (13 percent) for total organic carbon shown in Table 47 are somewhat suprising. Possible explanations for these low

TABLE 46.   EP TOXIC LEACHATE SUMMARY

Element	Concentration (mg/L)	EPA standard[a] (mg/L)
Arsenic	0.03	5.0
Barium	0.042	100.0
Cadmium	0.002	1.0
Chromium	0.065	5.0
Lead	0.17	5.0
Mercury	0.0004	0.2
Selenium	0.04	1.0
Silver	0.12	5.0

[a]U.S. EPA, Federal Register, Vol. 45,
No. 98: 33122. May 14, 1980.

TABLE 47.   FACILITY E TOTAL ORGANIC CARBON AND VOLATILE ANALYSIS

Parameter description	Samples ID and description		
	ACT-1-1 Contaminated solution	ACT-1-2 Contaminated solution- duplicate	ACT-1-3 Filtered solution	Average removal[a] (percent)
Total organic carbon (mg/L)	257.9	241.4	218.4	13
Volatiles (mg/L)				
Sulfur dioxide	2.1	$< 5 \times 10^{-6}$ [b]	$< 5 \times 10^{-6}$	99.998
Methyl formate	1.9	1.9	2.3	NC[c]
Methyl acetate	0.43	0.52	0.65	NC
Acetone	--	--	0.08	NC
Unknown	0.22	0.17	--	--
Unknown	0.12	0.10	--	--

[a] % removal $= \dfrac{\overline{ACT-1} - (ACT-1-3)}{\overline{ACT-1}} \times 100$

where: $\overline{ACT-1} = (ACT-1-1 + ACT-1-2)/2$.

[b] Detection limit for GC/FID analysis.

[c] NC = Not appropriate for calculation.

removals include the presence of activated carbon residuals in the filtered solution (ACT-1-3) or difficulties in analyzing the sample matrix. These results do not necessarily indicate that the brightener system was not preferentially adsorbed as shown below. For example, the high molecular weight polynuclear aromatics of the type present in the carrier component of the brightener system will readily adsorb to powdered carbon (EPA-600/8-80-023).

Conversely, low molecular weight carboxylic acid derivatives such as methyl formate, which do not effect plating quality, will not be easily adsorbed, especially in an acidic environment.

Volatiles and Semivolatiles--
The volatiles test results presented in Table 47 show little, if any, adsorption of low molecular weight carboxylic acid derivatives such as methyl formate, and methyl acetate. On the other hand, sulfur dioxide, which is thought to be a by-product of the brightener system, was completely removed. It is more likely that the sulfur dioxide was volatilized during treatment than adsorbed by the activated carbon. It must be remembered, however, that the adsorption of organic substances from mixed solution is a complex phenomenon. This can manifest itself in preferential adsorption of one substance over others, nonadsorption if a substance is only weakly adsorbed, or the displacement of a weakly adsorbed substance by a strongly adsorbed substance.

The semivolatiles test results, including nontarget compound analyses, indicated that all compounds that could be identified were below priority pollutant detection limits of 0.1 mg/L, except for phenols which were below the detection limit of 5.0 mg/L. It may be concluded then, that the spent plating bath and the spent activated carbon residue did not contain any priority pollutants, possibly making the residue suitable for delisting.

Cyclic Voltaic Stripping--
Cyclic voltaic stripping is an electrochemical analysis recently developed by Haak, Ogden, and Tench (Plating and Surface Finishing, September 1979) for the determination of brightener concentrations in acid copper baths. The method is one in which the potential of a rotating platinum disc electrode is cycled at a constant rate. Copper is alternately deposited on the electrode and stripped off by anodic dissolution. The resultant current density is plotted against the electrode potential to determine brightner concentration (Plating and Surface Finishing, December 1985). The analysis is able to determine brightner concentrations with 5 to 10 percent variance and a sensitivity of 0.2 mg/L on total brightner concentrations of approximately 5 mg/L.

Previously, with standard analytical techniques such as spectrometry or chromatography, it was difficult to control the concentration level due to the interference of other bath components. However, with the CVS analysis, the brightener concentration removal value could be easily determined and would be in direct proportion to the decomposition product removal rate. On May 23, 1986, a CVS analysis was performed by the SEL-REX division of the OMI Corporation. The brightener concentration was determined before and after activated carbon treatment (ACT-1-1 and ACT-3-1). Prior to activated carbon treatment the brightener concentration in the plating solution was 6.42 ml/L. After carbon adsorption the total brightener concentration was 3.40 ml/L, representing a 47 percent adsorption of brightener and byproducts.

ECONOMIC AND ENVIRONMENTAL EVALUATION

Economic Evaluation

Evaluation criteria for the processing of contaminated electrolytic plating baths for recovery and reuse include compliance with environmental regulations and overall economics. Regulatory justification is based on RCRA cradle-to-grave hazardous waste disposal responsibilities which include ultimate liability for the mismanagement of hazardous waste. Economic justification for the use of spent plating bath reclamation technology is related to the increasing costs of raw materials and regulatory compliance (waste treatment and disposal). A detailed current (i.e., 1986) cost estimate and economic evaluation is presented in Table 48.

Capital Costs--
Capital costs for the treatment system are based on a Baker Brothers Model 3020 Y activated carbon filtration unit with slurry tank. The stainless steel unit consists of 22 filter pads with a total available filtration area of 40.5 ft^2. Nominal capacity is 3,700 gpm, although at open pumping capacity, throughput is increased to 4,800 gpm. The equipment is delivered preassembled and FOB cost for one unit is $8,356. The capital cost estimate includes a contingency charge of 10 percent for onsite equipment modifications and related costs.

Operation and Maintenance Costs--
O&M costs are based on the operation of the activated carbon filtration unit for 8 hours per treatment, 28 treatments per year. This includes four activated carbon filtrations per year (one every 3 months) on each of the four, 2,400 gallon acid copper baths, and 12 (once/month) filtrations on the 1,200 gallon 60/40 tin-lead plating bath. Electricity costs are based on the operation of one 440V, 3 hp. TEFC motor required to recirculate contaminated solution throughout the treatment system. Labor costs were estimated only for the operation of the unit and consist of 8 labor hours per treatment, at 28 treatments/year. Treatment chemical costs consist of 88 lbs of powdered activated carbon and 4 gallons (47.4 lbs) of 50 percent reagent grade hydrogen peroxide per acid copper treatment. Each 60/40 tin-lead bath treatment was estimated to consume approximately half of these quantities due to differential bath volumes. MEFIAG filter papers are used at a rate of 22 paper-lined filters/pass, three passes/treatment. Replacement costs are $166/case of 250 filter papers. In addition to these operational costs, an annual maintenance charge of 10 percent of the total capital has been included.

Total Annual Costs--
Total annual costs for the implementation of the activated carbon filtration system in use at Facility E were approximately $10,153 and consist of total capital, operation and maintenance, and spent activated carbon disposal. The total capital cost was amortized over 10 years at 12 percent

TABLE 48.   ECONOMIC EVALUATION OF FACILITY E'S FILTER TREATMENT SYSTEM

Cost item	Unit cost ($)	Total Cost ($)
Capital Costs		
(1) Model 3020Y[a] Filter Treatment System	8,356	8,356
Miscellaneous	10% of purchase price	836
TOTAL CAPITAL		9,192
Annual O&M		
Mefiag Filter Papers[a]	166/250	1,227
Electricity[b]	0.05/KWH	25
Maintenance	10% of Total Capital	919
Labor	15/hr	3,360
Powdered Activated Carbon[c]	0.96/lb	1,859
50% Hydrogen Peroxide[c]	0.56/lb	583
TOTAL O&M		7,973
Annual Costs		
Annualized Capital[d]	0.177	1,627/yr
Annual O&M		7,973/yr
Annual Spent Carbon Disposal[e]	140/drum	553/yr
TOTAL COSTS		10,153/yr
Annual Credit		
Hazardous Waste Disposal[e]	1.15/gal	12,420/yr
Recovered Plating Solution[f]	(Copper) 10,000/bath	40,000/yr
	(Tin/Lead) 15,000/bath	15,000/yr
TOTAL CREDIT		67,420
TOTAL NET CREDIT (annual basis)		57,267

[a]Baker Brothers Technical Bulletin.

[b]Department of Energy, Energy Information Administration.   National Average, December 1986.

[c]McKesson Chemical Technical Brochure.

[d]Annual costs derived by using a capital factor:

$$CRF = \frac{i(1+i)^n}{(1+i)^{n-1}}$$

where: i = interest rate and n = life in the investment.   A CRF of 0.177 was used to prepare cost estimates in this document.   This corresponds to an annual interest rate of 12 percent and an equipment life of 10 years.

[e]As quoted by Clean Harbors Inc.

[f]OMI SEL-REX telecon.

interest. Annual spent activated carbon disposal costs (28 activated carbon treatments/year) are based on telephone conversations with several hazardous waste disposal companies. However, analytical test results seem to indicate that if the spent activated carbon is noncorrosive in nature, it may be suitable for delisting. Delisting would further minimize the quantity of hazardous waste generated at Facility E and decrease annual treatment costs by an additional 5 percent.

Total Annual Cost Savings--

Total annual savings for the implementation of the activated carbon filtration system were approximately $67,000 and consist of raw material purchase and hazardous waste disposal costs. The raw material purchase savings consist of 10,800 gallons of recovered plating solution at a cost of $10,000/acid copper bath and $15,000/solder bath. Recovery volumes are based on the assumption that prior to one full year of operation, organic impurity concentrations are noncritical. After this point (based on a high volume work flow), deposits will become burnt and powdery in nature and the bath will have to be replaced. Disposal costs are related to current (1986) hazardous waste facility pricing and represent a significant quantity of the total savings (18.4 percent). As evidenced from Table 48, approximately $57,267 of net savings were realized annually by utilizing the activated carbon filtration system. This represents a payback period of under 3 months for this application.

Environmental Evaluation

The reduction in quantity of hazardous waste that could possibly be land disposed is significant. Although most of the metals are reclaimed by either the manufacturer of the plating bath or a commercial hazardous waste treatment facility, significant quantities of hazardous waste treatment sludge (F006) are produced and are currently landfilled. Activated carbon treatment, therefore, is a cost-effective and environmentally sound technology for reducing the quantity of hazardous waste generated by electrolytic plating baths.

10. Facility F Case Study

Facility Description

Facility F manufactures primarily double-sided single-layer printed circuit boards using the subtractive method. The company is a job shop employing approximately 300 people and producing an average of 40,000 square feet/month of boards. Figure 12 details the process operations employed at Facility F.

Waste Sources

Approximately 75,000 gallons/day of metals contaminated rinsewaters are generated at Facility F. The rinsewaters which are of most concern, due to contamination with dissolved metals such as copper and lead, are shown (marked with an asterisk) in Figure 12. These rinsewaters generally contain very dilute concentrations of the process bath constituents. Standard bath constituents and their approximate concentrations (in the concentrated baths) are shown in Table 49. Another wastewater source is the rinses from photoresist developing and stripping operations. The photoresist used at this facility is developed using an aqueous solution containing sodium carbonate and butyl carbitol. Following the electroplating step, the light-exposed resist is stripped using a different aqueous solution containing glycol ethers and low and high molecular weight alcohols. Because both of these solutions are primarily water, they are combined with other wastewaters and then discharged to the sewer.

In addition to rinsewaters and resist developing and stripping solutions, there are several process baths that are sent offsite after use to be regenerated or disposed of. These include spent copper etching solutions, spent solder stripping solutions, solutions used to strip metal from plating racks, and spent solvent used to strip epoxy inks.

The copper etchant solution is an aqueous, alkaline solution containing 12 percent ammonium chloride. After a period of use, it will accumulate up to 15 percent dissolved copper. At this point, it will have lost its effectiveness and must be replaced with fresh etchant. The spent etchant is placed in a 4,500-gallon storage tank where it awaits pickup and regeneration by the manufacturer. Approximately 70,000 gallons are sent offsite each year.

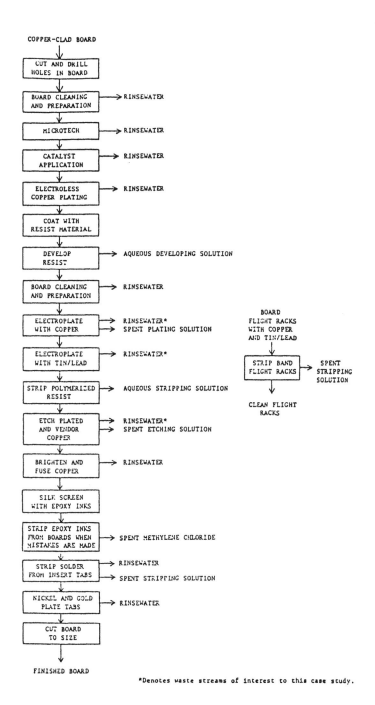

Figure 12.   Workflow diagram for Facility F.

TABLE 49. COMPOSITION OF PROCESS BATHS

Process bath	Major constituents	Concentration (g/L)
Acid copper[a,b]	Sulfuric acid	52.5–135
	Copper sulfate	160–300
	Copper (Cu^{+2})	40–75
	PC gleam (brightening agent)	--
	Chloride ion	20–80 ppm
Solder bath (60% Tin – 40% Lead)[a,b]	Stannous tin	56.2
	Lead	26.2
	Fluoboric acid	100.0
	Boric acid	26.2
	Peptone	5.2
Copper etchant[c]	Ammonium Chloride	12%
Tin/lead strip[c]	Hydrogen Peroxide	10%
	Fluoric Acid	20%
Copper microetch[c]	Sulfuric Acid	--
	Hydrogen Peroxide	--
Electroless copper[d]	Copper	1.5–2.4 g/L
	Formaldehyde	1.5–3.0
	NaOH	6–8
	chelating agents	--

[a]EPA-600/2-83-033.

[b]Baths which contaminate rinses that were tested in this study.

[c]From plant-supplied Material Safety Data Sheets.

[d]Product literature from Shipley Company, Inc.

The solder stripping solution is also regenerated offsite by the manufacturer. It is composed of 20 percent fluoric acid and 10 percent hydrogen peroxide, and will have accumulated high concentrations of tin and lead after use. Approximately 30, 55-gallon drums are generated per month.

The only waste stream which is actually disposed of, as opposed to being regenerated, is the spent rack stripping solution. These racks hold the boards as they are immersed in the copper and the tin/lead plating solutions, and consequently they also are plated with metal. Periodically, the racks are placed in a solution composed of 50 to 70 percent nitric acid to remove the plated metal from the stainless steel rack. The spent solution, containing dissolved metals, is then picked up for offsite disposal as a hazardous waste. Approximately 8,000 gallons of spent solution are generated annually.

Finally, a waste stream is generated by stripping epoxy inks from circuit boards using methylene chloride. These inks are applied prior to gold plating the tabs by silk screening through a mask with the image of the circuit pattern on it. If a mistake is made due to misalignment of the mask, the epoxy ink with methylene chloride is removed prior to curing. The spent methylene chloride is sent offsite to be reclaimed.

Waste Management

Background--

The offsite management of several waste streams was mentioned above. Of concern to this study, however, are onsite methods of reducing the quantities of waste that would otherwise be managed offsite. At this facility, the major process of this type is electrolytic recovery of metals from rinsewaters. Electrolytic recovery has been practiced on rinses following several plating baths for a little more than 1-year. The primary purpose of the electrolytic reactors is to reduce the concentration of metals in rinsewaters which are released to the wastewater sump. Prior to recovering metals from these rinsewaters, a simple two-stage rinse system was used. This resulted in the release of up to 3,000 ppm of copper and lead to the sump which would have necessitated some type of end-of-pipe treatment system in order to comply with increasingly strict pretreatment standards. Instead of installing an end-of-pipe treatment system, however, a decision was made to try to attain compliance by removing the contaminants at the source. Not only would this be a much less expensive alternative, but it would also eliminate the generation of large quantities of hazardous sludge that are associated with most conventional treatment systems.

The location of the electrolytic recovery units has been changed several times to achieve the greatest recovery of metals. For example, originally there was one unit used to recover copper from the rinse following electroless copper plating. The amount of copper recovered, however, was low and so the unit was moved to the copper electroplating rinse where there are higher concentrations of dissolved copper with potential for recovery. There are now four individual copper recovery units associated with this process, and there are three units associated with the solder (tin/lead) electroplating process.

Currently, there are no longer recovery units associated with the copper etching rinses, the third major source of metal-containing rinsewater. This is because the electrolytic units which are currently in use do not have the capability to recover copper from etching solutions. Electrolytic recovery from copper etching solutions is difficult because the purpose of the etching solution is to remove plated copper, and so once the copper is plated onto the cathode of the electrolytic cell, it is quickly etched back into the solution. Facility F is currently investigating the use of more powerful units to recover copper from this waste stream.

Electrolytic Recovery System--

The installation of electrolytic recovery units required converting the primary rinse tank into a static dragout tank, as shown in Figure 13, and leaving the second rinse tank as a flowing rinse. The contents of the dragout tank are continuously circulated through the electrolytic reactor(s) and back into the dragout tank. As the solution passes through the reactor a small amount of metal is plated onto the cathode. Since plating solution, containing dissolved metals, is continuously input to the dragout tank, the removal of metal by the electrolytic reactor is only able to maintain a certain concentration of metals in this solution. The concentration is maintained, however, at a low enough level so that drag-in of metals to the secondary rinse is minimal. The secondary rinse solution can then be released to the wastewater sump containing only a low concentration of metals.

The electrolytic reactors used at this facility are simple, compact units. They consist of a wastewater sump, a pump, and the anode and cathode, contained within a rectangular box with dimensions of approximately 22 in. x 10 in. x 22 in. The anode is cylindrical and is encircled by a stainless steel cathode with a diameter of 8 inches and a height of 6 inches [Agmet Equipment Corp.]. The anode material used for copper plating solutions is titanium. For tin/lead plating solutions, however, the anode material is columbium. The columbium anode is required for the tin/lead rinse because the fluoroboric acid in these solutions was found to be extremely corrosive to titanium. Other pertinent characteristics of the electrolytic reactors tested are presented in Table 50. These units are operated at constant voltage, and so the amperage will vary according to the conductivity of the solution. The higher the concentration of electrolyte in the dragout tank, the higher will be the corresponding current.

TABLE 50.   ELECTROLYTIC REACTOR CHARACTERISTICS

Design parameter	Cylindrical anode within cylindrical cathode
Cathode area	1 ft^2
Reactor volume	1.3 ft^3
Maximum flowrate	16.3 gallons/minute
Amperage	0.5 to 20 amps
Power	110 VAC
Cathode material	Stainless steel
Anode material	Columbium or titanium

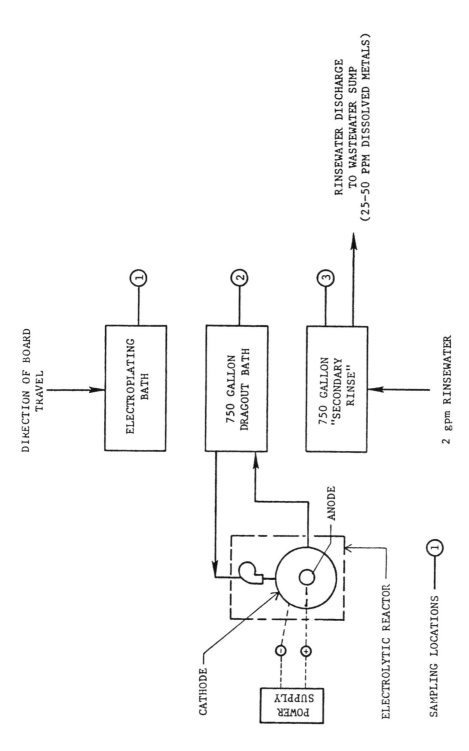

Figure 13.   Electrolytic recovery units employed at Facility F.

The electrolytic units are operated 24 hrs/day, 7 days/week, except for period maintenance requirements. Maintenance includes replacing parts, especially related to the pump mechanism, and removing the metal foil that has been plated onto the cathode. Foil removal is usually necessary once a week for each of the units. The time required to clean each unit is about a 1/2-hour. Copper recovery per week has averaged about 10 pounds, and lead recovery has averaged about 5 pounds/week.

PROCESS TESTING AND ANALYTICAL RESULTS

Process Testing

Sampling of rinsewater and process streams associated with copper and tin/lead electroplating was conducted on February 18-19, 1986. There were three sampling locations associated with each of the electroplating processes. These were the plating bath, the dragout bath and the secondary rinse. The sampling activities were conducted over a 24-hour period starting and ending at approximately 9:00 a.m. The most important parameter to define was the concentration of dissolved metals (copper, lead and tin) in the dragout bath. Since the dragout bath is circulated through the electrolytic reactor 24 hours/day, samples were taken every 4 hours over the 24-hour period. The printed circuit board plating line is only operated for 16 hours out of the 24-hour period (between 8:00 a.m. and midnight) and, therefore, samples of the secondary rinse bath were taken every four hours during this period.

Samples were also collected for analysis of total organic carbon (TOC). These samples were collected to provide a general indication of the fate of organic compounds when subjected to the electrolytic reactor. Because lesser importance was attached to these types of compounds, sample collection was less frequent.

Operation on the day of testing deviated from normal due to several factors. The first of these was that the dragout tank following copper electroplating had been emptied the previous day in order to fix the weir. It was then refilled with fresh water. Consequently, the concentration of metals in the tank was not as high as it normally would be. In addition to this, one of the electrolytic units stopped operating during the middle of the testing because of a broken pump impeller. The effect of this would be to reduce overall metal recovery from the bath 25 percent (since there are four units altogether), and thus result in higher concentrations of metal in the dragout bath and the secondary rinse tank.

Finally, the cathodes of one tin/lead and one copper recovery unit were weighed at the beginning and at the end of the 24-hour sampling period in order to determine the quantity of metal that was recovered from solution. For the copper recovery unit, the weight increase was 0.30 pounds. For the tin/lead unit there was no difference in the weight of the cathode at the beginning and the end of the 24-hour period, indicating that there were problems with the unit.

Analytical Results

Copper Electroplating--
    Table 51 presents the measured concentrations of copper, tin, lead, and total organic carbon in the copper electroplating, dragout and rinse baths. The concentrations of these constituents, particularly copper were measured to determine the performance of the electrolytic reactors with regard to removing metals from the dragout bath, and as a result reducing the concentration of metals in the secondary rinse bath.   Several indicators of reactor performance are discussed below.

    Secondary Rinse Copper Concentrations--The concentration of copper or other constituent in this stream is important to know because this is the stream that is actually released to the sewer (after mixing with other wastewater streams).   As listed in Table 51, the concentration of copper ranged from 70 to 90 mg/L over the 24-hour period with the highest concentration being at 8:00 a.m.   The contribution of 90 mg/L of copper at 2 gpm to the final plant effluent can be estimated assuming the total plant effluent is 75,000 gallons/day.   This calculation is shown below:

    90.4 mg/L x 2 gal/min x 960 min/day / 75,000 gal/day = 2.31 mg/L

The maximum allowable daily discharge of copper is 4.5 mg/L (40 CFR 413). Therefore, the other sources of copper must be kept below 2.2 mg/L for these discharge limits to be met.   Since there are several other sources of copper, particularly that from the rinse following copper etching, these limits may be difficult to achieve without increasing the number or power of the electrolytic reactors.

    Copper Recovery Rate--The rate of recovery of copper by the electrolytic reactors can be determined in several ways.   One of these ways is to monitor the concentration of copper in the dragout tank over time.   A plot of this relationship is shown in Figure 14.   This plot shows that the copper concentration increases in an approximately linear fashion between 1:00 p.m. and midnight at a rate of 21.5 mg/L/hour.   This increase in copper concentration occurs despite the removal of copper by the electrolytic reactors, indicating that the rate of input of copper due to dragout from the plating bath is greater than the rate of removal achieved by the electrolytic reactors.

    From midnight until 9:00 a.m., when there is no dragout of copper from the plating bath, the copper concentration decreases at a rate of approximately 5 mg/L/hour.   Knowing that the size of the dragout tank is 1,000 gallons, the mass rate of removal of copper is calculated to be equivalent to almost 19 grams/hour.

    The rate at which copper is removed from the dragout tank was also estimated by weighing the cathode of one of the electrolytic reactors both at the beginning and the end of the 24-hour testing period.   This weight measurement showed a 0.30 lb increase over the 24-hour period indicating that an average of 5.7 grams/hour of copper were removed from solution and plated onto the cathode.   Since there are four electrolytic units used on the copper

TABLE 51.   CONSTITUENT CONCENTRATIONS FOR COPPER ELECTROPLATING PROCESS (mg/L)

Time	Plating bath				Dragout tank				Second rinse		
	Cu	Pb	Sn[a]	TOC	Cu	Pb	Sn[a]	TOC	Cu	Pb	Sn[a]
9:00 am	27,000	3.2	17	873.1				11.42			
1:00 pm					326	0.45	3.2	14.05	70.5	0.26	4.0
5:00 pm					416	0.54	2.8	16.90	77.6	0.08	4.3
9:00 pm					498	0.38	5.1		83.6	0.16	5.3
1:00 am					530	0.64	3.2				
5:00 am					508	0.21	3.7				
8:00 am				888.6				23.13	90.4	0.35	5.0
9:00 am					490	0.45	3.8				

[a]Tin data considered invalid, see Appendix A.

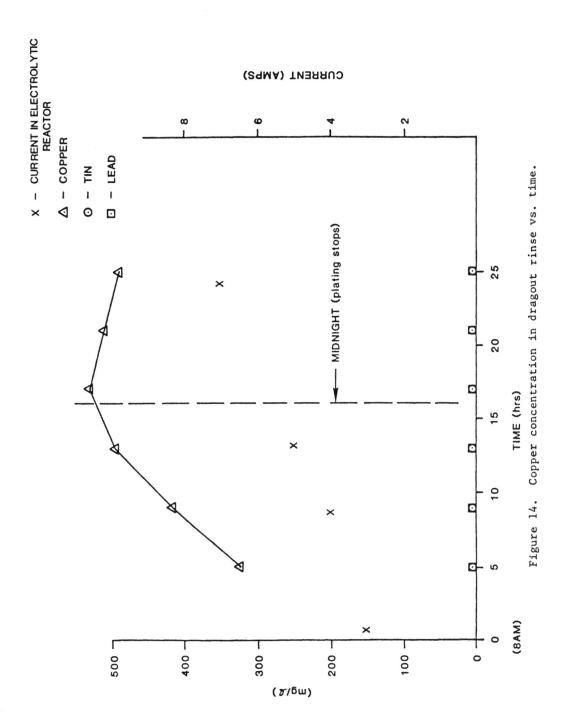

Figure 14.   Copper concentration in dragout rinse vs. time.

dragout tank, the total amount of copper removed would be approximately 22.7 grams/hour. During testing, however, one of the units broke down and did not function for the full 24-hour period. Therefore, the quantity of copper removed would be slightly less than expected. Nonetheless, this method and the graphical method of estimating copper removal rates yield similar results.

Finally, one can determine the efficiency of the reactors by comparing the actual amounts of copper recovered to the theoretical maximum amount that could be removed. Faraday's Law states that the amount of material that can be produced electrochemically is proportional to the amount of charge in coulombs [Snoeyink, V. L. and D. Jenkins, 1980]. A coulomb is equivalent to the amount of charge transferred when 1-ampere of current flows for 1-second. Using this fact and knowing that the average current over the 24-hour period was 5 amperes (see Figure 14), one can calculate the theoretical amount of copper recovered in 1-hour.

$$\left(\frac{31.7 \text{ g Cu}}{\text{equivalent}}\right) \times \left(5 \text{ amperes}\right) \times \left(3,600 \text{ sec/hr}\right) \bigg/ \left(96,500 \frac{\text{coulombs}}{\text{equivalent}}\right) = 5.9 \text{ g/hr}$$

This is the amount of copper recovered per reactor, so with four reactors the amount that theoretically could be recovered is 23.7 grams/hr. Then, the current efficiency of the reactors can be determined by dividing the estimated "actual" copper removal rate into this theoretical rate. Depending on which estimate of actual removal rate is used (18.9 or 22.7) the current efficiency is calculated to be between 80 and 90 percent. As indicated in Table 52, however, the removal efficiency of the electrolytic reactors, based on a copper input rate of 100.3 grams/hour is between 18 and 22 percent.

TABLE 52. COPPER RECOVERY DATA

Copper input to dragout tank[a]	100.3 grams/hour
Copper removal rate	18.9 - 22.7 grams/hour
Theoretical recovery rate	23.7 grams/hour
Current efficiency	80 - 96 percent
Removal efficiency	18 - 22 percent

[a]Calculated by assuming that total copper input is equal to the rate of increase of copper concentration between 8:00 a.m. and midnight, plus the rate of decrease of copper concentration between midnight and 8:00 a.m.

Tin/Lead Electroplating--

Table 53 presents the measured concentrations of tin, lead and copper in the tin/lead electroplating dragout and rinse baths. Due to the complexity of the sample matrix, the accuracy of the tin analyses on all baths and the analyses of lead in the plating bath were very poor. As a result, it was not possible to use these data to make any definitive conclusions. In addition, as is discussed below, the data on the tin/lead baths in general does not show a

TABLE 53.   CONSTITUENT CONCENTRATIONS FOR TIN/LEAD PLATING PROCESS (mg/L)

Time	Plating bath			Dragout tank			Second rinse		
	Cu	Pb	Sn[a]	Cu	Pb	Sn[a]	Cu	Pb	Sn[a]
9:15 am	2.91	5,000[a]	2,400						
1:15 pm				4.88	2,400	2,200	1.57	45	38
5:15 pm				4.76	2,500	2,100	0.43	56	11
9:15 pm				4.28	2,100	380	0.38	64	13
1:15 am				3.36	2,500	410			
5:15 am				4.22	2,300	1,500			
8:15 am	3.02	5,700[a]	2,700				0.28	22	9.0
9:15 am				4.23	1,900	1,700			

[a]Data considered invalid, see Appendix A.

clear relationship between metal addition and removal as did the copper data.
This may be in part due to the analytical difficulties, but it may also
indicate that the electrolytic reactors were not functioning properly.
Nonetheless, it was possible to make some general conclusions. These are
discussed below.

Plating Bath--The average, measured concentrations of tin and lead in the
tin/lead plating bath are, respectively, 2,550 and 5,350 mg/L. These
concentrations are almost an order of magnitude lower than the concentration
of copper in the copper electroplating bath. Consequently, the amount of
metal which will be dragged out of the bath should also be lower.

Dragout Bath--The measured concentrations of lead in the dragout bath
range from 1,900 to 2,500 mg/L. These concentrations are at most 85 percent
less than the concentration of the metal in the plating bath itself, and more
commonly the concentration is only 50 percent less or close to equivalent.
This indicates that the lead is only being removed to a very small degree by
the electrolytic reactors.

Figure 15 shows the tin/lead curve for metal concentration vs. time. In
contrast to the copper plating case; this data does not show a clear
relationship between input of metal to the dragout bath and removal by the
electrolytic reactor. Instead, both the tin and lead concentration appear to
drop during the period when it would be expected that the input of metal would
exceed the removal by electrolytic recovery. Then, during the midnight to
8:00 a.m. period, when there is no input of metal to the dragout bath, the tin
concentration rises from less than 500 mg/L to greater than 1,500 mg/L. As
mentioned above, the difficulties in analyzing these samples may be the cause
of these unexplainable results.

Two other indicators of poor recovery of lead and tin from this dragout
bath are:

- The low amperage of the electrolytic reactors; and

- The unmeasurable amount of metal plated onto the cathode of one of
the reactors.

Firstly, the current indicated by the ammeter on the electrolytic units
remained below 1-ampere during the entire test period. With a current of
1-ampere, the maximum removal of tin and lead would be, respectively, 1.1 and
1.9 grams/hour/unit (from Faraday's Law). To increase the rate of recovery,
the amperage and/or the number or electrolytic recovery cells could be
increased. Increasing the amperage, however, may increase the generation of
gases (such as oxygen and fluorine) at the anode and result in the plated
metal being of poorer quality. This may then result in the etching of plated
metal back into solution.

The other method for quantifying the removal of metals by the
electrolytic reactor was to weigh the cathode of one of the units at the
beginning and the end of the 24-hour period. In doing this, the weight of the
cathode did not change, indicating that no metal was removed from solution by
this electrolytic reactor.

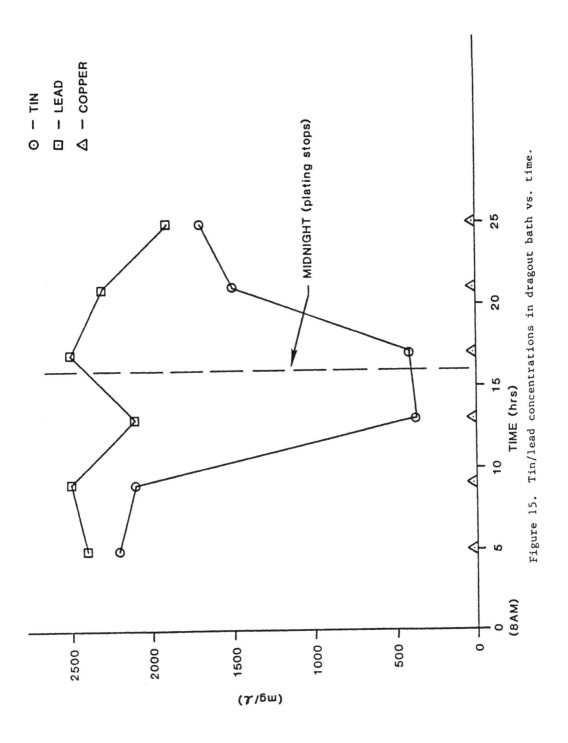

Figure 15. Tin/lead concentrations in dragout bath vs. time.

Second Rinse--The concentrations of metals in the secondary rinse are also erratic and do not clearly correspond to the concentrations in the dragout bath. This may be due to analytical difficulties with these samples. Nonetheless, averaging the four analyses for each metal results in a lead concentration of 47 mg/L. Assuming a total plant effluent of 75,000 gal/day, the concentration of lead in the final effluent due to this 47 mg/L would be 1.2 mg/L. Maximum daily allowable levels of lead are 0.6 mg/L. In order to achieve this level, the concentration of lead in the secondary rinse would have to be lowered from 47 to 23 mg/L.

ECONOMIC AND ENVIRONMENTAL EVALUATION

This facility installed electrolytic recovery units to reduce the amount of metals in its final effluent. It appears, however, that this technology is not one that can be used by itself to achieve effluent discharge limits, particularly for lead. Its advantage, then, is to remove some of the metals from the rinsewaters at the source, thereby lessening the amount of end-of-pipe treatment that must be done. For example, if a precipitation system is used to treat the total plant effluent, the amount of metal hydroxide sludge that will be generated by this system can be reduced by removing some of the metals upstream using electrolytic recovery. Reducing the quantity of sludge that is generated will be beneficial in both economic and environmental terms.

Economic Evaluation

Tables 54 and 55 present cost estimates for electrolytic recovery of metals from dragout baths following copper and tin/lead plating. Separate estimates are presented for the copper recovery and the tin/lead recovery systems because the equipment cost for the two systems is different, and also because maintenance requirements for the tin/lead system are expected to be higher than for the copper system. Both of these cost differences are due to the tin/lead bath being extremely corrosive, therefore requiring corrosion-resistant columbium anodes, and more frequent maintenance to replace corroded parts. The basis for other elements of these cost estimates is discussed below.

Capital Cost--
As mentioned above, the purchase cost for one unit to recover copper from an acid copper rinse is less than the cost of one unit to recover lead from a tin/lead fluoborate bath. The difference of 1,000 dollars, 3,500 vs. 4,500 dollars, is due to the use of a columbium vs. a titanium anode. Otherwise, the two units are identical. The cost quoted here is the cost for which these units (Agmet Model 5200) were purchased in 1985.

The other element included in the capital cost is for miscellaneous items associated with the installation of the unit. These include piping and any changes that have to be made to the rinse baths in order to install the units. Miscellaneous costs have been set at 10 percent of the equipment cost.

TABLE 54.   ELECTROLYTIC COPPER RECOVERY

Basis	Unit cost ($)	Cost ($)
Capital costs		
• 4 recovery units with Titanium anode[a]	3,500[a]	14,000
• Miscellaneous costs such as installation/piping	10% of total capital costs	1,400
Total capital		15,400
Operation & maintenance		

• Electricity
  - for electrolysis (3 volts, 10 amps/unit)
  - for pumps (1/8 HP/unit)
	$0.05/kwhr[b]	45
		139

• Maintenance	10% of capital costs	1,540
• Labor (250 hrs/yr)	$15/hr	3,750
Total O & M		5,474
Annual costs		
• Annualized capital (10% over 10yrs)	0.1627	2,505
• O & M		5,474
Total annual cost		7,980
Annual savings		
• Recovered copper (10 lbs/wk)	$0.22/lb[c]	114
• Sludge disposal 2.3 tons at 20% solids or 23 tons at 2% solids	$200/ton	455-4550
• Waste treatment chemicals (0.4 tons Ca(OH)$_2$)	$50/ton[d]	20
Total annual savings		589-4684

[a]Agmet Equipment Corporation.

[b]Department of Energy, Energy Information Administration. National Average. December 1986.

[c]Price quoted to Facility F.

[d]Chemical Marketing Reporter. May 5, 1986.

TABLE 55.   ELECTROLYTIC TIN/LEAD RINSE RECOVERY

Basis	Unit cost ($)	Cost ($)
Capital costs		
• 3 recovery units with columbium anode	4,500	13,500
• Miscellaneous costs	10% of capital costs	1,350
Total capital		14,850
Operation & maintenance		
• Electricity	$0.05/kwhr	
- for electrolysis		35
- for pumps		104
• Maintenance	20% of total capital costs	2,970
• Labor (250 hrs)	$15/hr	3,750
Total O & M		6,859
Annual costs		
• Annualized capital (10% over 10 yrs)	0.1627	2,416
• O & M		6,859
Total annual cost		9,275
Annual savings		
• Sludge disposal 0.9 tons at 20% solids or 9 tons at 2% solids	$200/ton	185-1,850
• Waste treatment chemicals (0.2 tons Ca(OH)$_2$)	$50/ton	10
Recovered tin/lead (5 lbs/wk)	$0.10/lb[a]	26
Total savings		221-1,886

[a]Price quoted to Facility F.

Operation and Maintenance Costs--
    The first of the costs listed under this heading is electricity costs.
Electricity is required for the 1/8-horsepower pump contained within each
electrolytic reactor and also for generating the electric current necessary to
plate the metal onto the cathode. The electricity use for copper recovery is
based on operation at 3 volts and 10 amperes, and for tin/lead the recovery
voltage is 6 and the amperage is 5.

    Maintenance costs for the tin/lead units are assumed to be 20 percent of
capital costs vs. 10 percent of capital costs for the copper units. As
mentioned above, this is due to the anticipated higher frequency of parts
replacement resulting from the highly corrosive nature of the fluoroboric acid
in the tin/lead plating solution. Personnel at this facility have indicated
that proper maintenance of the units is extremely important in order to
achieve maximum recovery.

    Labor associated with these units is primarily for removing the plated
metal foil from the cathode of each of the units. This must be done
approximately once per week for each unit. Labor is also required for fixing
units that have broken down. Approximately 5 hours/week for the copper units
and the tin/lead units is assumed to be required.

Annual Costs--
    Annualized capital cost was estimated using an interest rate of
10 percent over a 10-year period. Assuming a rate of recovery of 10 lbs of
copper/week, the total annual costs/pound of copper recovered would be
approximately 15-dollars. For tin/lead recovery of 5 lbs/week, the cost would
be almost 30 dollars/pound.

Annual Savings--
    The use of electrolytic recovery units to remove metals from rinsewaters
at the source of generation will lower the amount of metals that must be
removed in an end-of-pipe treatment system. The "savings" that are presented
at the bottom of the table are those that would be accrued if electrolytic
reactors were used upstream of an end-of-pipe lime precipitation system. When
the electrolytic reactors are used, less metals reach the precipitation
system, and so less lime is required and less hydroxide sludge is produced.
The savings are based on recovery of 10 lbs/week of copper and 5 lbs/week of a
1:2 tin/lead mixture. The quantity of sludge not generated as a result of
recovering these metals would vary in volume depending on whether it was
thickened and dewatered. A range of cost values, one based on 20 percent
solids and the other based on 2 percent solids, is presented. A sludge of
20 percent solids is ten times less voluminous than one of 2 percent solids
and, therefore, the cost for disposing it would be correspondingly lower.
However, some type of equipment for dewatering the sludge, most likely a plate
and frame filter press, would be required to achieve 20 percent solids.
Therefore, the decreased sludge disposal costs would be offset by increased
equipment costs.

    Tables 54 and 55 indicate that the annual costs associated with
electrolytic recovery exceed the annual savings. The savings, however, are
based on sludge disposal costs of 200-dollars/ton. With upcoming land

disposal restrictions on certain metal-bearing hydroxide sludges, however, their disposal will most likely become much more expensive. Therefore, in the near future, the cost savings may become much greater.

Environmental Evaluation

The environmental benefit of electrolytic recovery is that the quantity of metal hydroxide sludge (RCRA code F006) that is generated by an end-of-pipe treatment system is minimized. The removal of 10 lbs/week of copper and 5 lbs/week of tin and lead from dragout rinse baths reduces by 32 tons/year the quantity of sludge (at 2 percent solids) that would otherwise be generated by precipitation. Instead of being converted to hydroxide sludges, the copper, tin and lead are plated onto the cathode in a metallic form so that they can be reused.

References

Agmet Equipment Corporation. Product Literature on Met-tronic 5200.
Cranston, Rhode Island

Aldrich, Roberts. Handel, Centec Corporation. Hazardous Sludge Reduction.
The 70th AES Annual Technical Conference (June 1983)
Proceedings, Indianapolis, IN

American Conference of Governmental Industrial Hygeinists (ACGIH).
TLV's for Chemical Substances in the Work Environment adopted by
ACGIH with intended changes for 1985-1986.
ISBN: 0.936712-61-9, 1985.

APV Crepaco, Inc. Price Quotation #D-9585-RG. March 24, 1986.

Baker Brothers Corporation. Product Literature on Model 3020 SY Carbon
Filtration Unit.

Baron-Blakeslee, Inc. Telecon with M. Arienti, GCA Technology Division, Inc.
March 28, 1986.

Blodgett, W.A. Assessment of Solvent Distillation Equipment. NEESA
20..3-013 Naval Energy and Environmental Support Service. December 1985.

Card, D. Flat Year Forecast for Independent PCB Markers. Electronic Business,
September 1, 1985, pp. 112-132.

Chemical Marketing Reporter. May 5, 1986.

Chillingworth, et al. Volume 1V of Industrial Waste Management Alternatives
and Their Associated Technologies/Processes. GCA-TR-80-80-G.
Section 5. pp. 40-48.

Clean Harbors Incorporated, Kingston, MA. Vendor quote for spent activated
carbon disposal, July 25, 1986.

Conoby, J.F. Circuit Board Facility Emphasizes Water Recovery and Reuse.
Plating and Surface Finishing. April 1984. pp. 38-40.

Conway, R.A., and R.D. Ross. Handbook of Industrial Waste Disposal.
1980 Edition. Van Nos Reinhold. pp. 213-215.

Department of Energy, Energy Information Administration. National Average.
December 1986.

Dietz, J.D. and Cherniak, C.M., University of Central Florida, Orlando. Evaluation of New and Emerging Technologies in the Metal Finishing Industry. U.S. EPA Water Engineering Research Laboratory, Cincinnatti, Ohio. December 1984.

Electronic Business. Top Semi Companies: A Changing of the Guard. January 1, 1986 pp. 78-80.

Electronic Business. A Semi Tough Scene for the Chip Kings. March 1, 1985 pp. 138-140.

Engineering Science. Supplemental Report on the Technical Assessment of Treatment Alternatives for Waste Solvents. Contract No. 68-03-3149. Final Report for EPA Office of Solid Waste. 1984.

Environmental Regulations and Technology. The Electroplating Industry. EPA/625/10-85/001.

EPA-600/8-80-023. Carbon Adsorption Isotherms for Toxic Organics. April 1980.

EPA-600/2-83-033. Industrial Process Profiles for Environmental Use: Chapter 36. The Electronic Component Manufacturing Industry. U.S. EPA April 1983.

EPA-625/5-85/016. Environmental Pollution Control Alternatives: Reducing Water Pollution Control Costs in the Electroplating Industry. September 1985.

Flick, E.W. Industrial Solvents Handbook. Third Edition, Noyes Data Corporation. Park Ridge, N.J. 1985.

Heleba, S.F. EPA Effluent Compliance and Sludge Control. PC FAB, May 1984 pp. 58-61.

Lindsay and Hackman. Morton Thiokol Inc. Sodium Borohydride Reduces Hazardous Waste. Purdue Research Foundation, West Lafayette, IN 47701. 1985.

Lopez, N. Chelated Copper Extricated by Membrane Filtration System, Effluent Cut to Less than 0.2 ppm. Chemical Processing. October 1984. pp. 94-95.

Lopez, N. Fair Treatment. Circuits Manufacturing. September 1984. pp. 116-118.

McKesson Chemical Company, San Francisco, CA. Technical brochure, June 1985.

Memteck Corporation, Woburn, MA. Technical brochure, November 1985.

Nemec, M.M. Zerpa Offers On-Site Solution for Solvents. Hazardous Materials and Waste Management, Nov./Dec. 1984.

OMl International Corporation, Nutley, New Jersey. Alliance telecon, July 25, 1986.

Pace Inc., Solvent Recovery in the United States 1980-1990.   Houston, TX.
    Prepared for Harding Lawson Associates, January 1983.

Patterson, W.P. "Gloom to Boom in Silicon Valley?", Industry Week,
    October 14, 1985.

PEDCo Environmental Inc. Industrial Process Profiles for Environmental Use:
    Chapter 36. The Electronic Component Manufacturing Industry.
    EPA-600-2-83-033. U.S. EPA. April 1983.

Peters, M.S. and K.D. Timmerhaus. Plant Design and Economics for
    Chemical Engineers. McGraw-Hill Third Edition. 1980.

Plating and Surface Finishing. Cyclic Voltammetric Determination of
    Brightner Concentration in Acid Copper Sulfate Plating Baths,
    December 1985.

Recyclene Products Inc. Technical Product Bulletin, 405 Eccles Avenue
    So. San Francisco, CA 94080. 1985.

Removal and Recovery of Heavy and Precious Metal with Sodium Borohydride.
    Metal R&R Newsletter. Issue No. 4. Morton Thiokol, Inc. 1985.

Shipley Company Inc. Product Literature on Cuposit CP-78. Newton, MA.

Snoeyink, C.L. and D. Jenkins. Water Chemistry. John Wiley and Sons, Inc.
    1980.

Ulman, J.A. Control of Heavy Metal Discharge in the Printed Circuit Industry
    with Sodium Borohydride. 1984 AES SUR/FIN Annual Technical Conference
    and Exhibit.

Ventron Corporation, Andover, MA. Technical brochure, November 1985.

Versar, Inc. National Profiles Report for Recycling--A Preliminary Assessment.
    Contract No. 68-01-7053. Draft Report for EPA Waste Treatment Branch.
    1985.

WAPORA Inc. Assessment of Industrial Hazardous Waste Practices - Electronic
    Components Manufacturing Industry, EPA SW-140C, January 1977.

Weast, R.C., Editor. CRC Handbook of Chemistry and Physics. 58th Edition.
    CRC Press, Inc., 2255 Palm Beach Lakes Blvd., West Palm Beach, Florida.

Wing, R.E. Complexed and Chelated Copper-containing Rinsewaters. Plating
    and Surface Finishing. July 1986. pp. 20-22.

Wopschall, R.M. High-Density Yield PWB Imaging with Dry Film Photoresist.
    Circuits Manufacturing, July 1984, pp. 36-40.

Yeshe, P. Low-Volume, Wet-Scrap Processing. Chemical Engineering Progress,
    September 1984, pp. 33-36.

Appendix A: Quality Assurance Summary

INTRODUCTION

Quality Assurance/Quality Control (QA/QC) procedures followed in this program were based upon routine laboratory and field practice and the Quality Assurance Project Plans prepared for this program in December, 1985. This Quality Assurance section will summarize areas where changes in laboratory and/or field procedures were made, and will address EPA comments on the Project Plan made in memoranda dated February 28, 1986. To facilitate review of pertinent QC data, this section will follow the outline of the QA Plan. For a detailed description of QA/QC data or procedures for each facility refer to either the Draft Report and/or the QA Plan for each facility.

PROJECT ORGANIZATION AND RESPONSIBILITY

During the course of this program, several major organizational changes were made. In the analytical laboratory, Dr. Peter Lieberman replaced Ms. Mary Kozik as Inorganic Section Head, and Ms. Joan Schlosstein replaced Ms. Andrea Cutter as Analytical QC coordinator. In the field measurements department, Mr. Howard Schiff replaced Mr. Graziano as Field QC coordinator.

PRECISION, ACCURACY, COMPLETENESS, REPRESENTATIVENESS AND COMPARABILITY

Analytical precision was estimated through the analysis of replicate sample aliquots. Analytical accuracy was determined through the analysis of EPA Environmental Monitoring and Support Laboratory (EMSL) Quality Control Samples and the analyses of matrix spiked sample aliquots. Results of these analyses broken down by facility are discussed below. Wherever possible, reference methods and standard sampling procedures were used as stated in the QA Plan to ensure comparability with other representative measurements made by Alliance or another organization.

Facility A

Quality control procedures for trace metals, total organic carbon, and total organic halide determination included the preparation and analysis of a laboratory method blank, for which final results were corrected, a laboratory control sample (LCS), duplicate sample aliquots and matrix spikes of duplicate aliquots. Laboratory control samples were obtained from U.S. EPA Environmental Monitoring and Support Laboratory-Cincinnati and prepared as directed to check instrument calibration. Results which are presented in

Table A-1 indicate that precision ($\leq$ 30 relative percent difference), accuracy (70-130 percent recovery), and completeness (95 percent valid) goals were met for the liquid wastes matrix determinations with the exception of total organic halides which had only 25 percent accuracy. Precision ($\leq$50 relative percent difference), accuracy (50-150 percent recovery) and completeness (95 percent valid) goals were met only for trace metals and total organic halides for the solid wastes matrix determination. Precision and accuracy goals for total organic halides in the solid wastes matrix was not performed.

Quality control procedures for the determination of total cyanides and hexavalent chromium included the preparation and analysis of a laboratory method blank by which final results were corrected, a laboratory control sample, duplicate sample aliquots and matrix spikes. However, due to the presence of interfering substances such as excess chlorine from the cyanide oxidation process and distillable organics precision and accuracy goals for total cyanide and hexavalent chrome were inconclusive or met in only a few cases. Completeness was judged to be 0 for these analyses.

Facility B

Quality control procedures for trace metals, total organic carbon, and total organic halide determination included the preparation and analysis of a laboratory method blank, by which final results were corrected, a laboratory control sample, a duplicate sample aliquot and a matrix spike of duplicate sample aliquots. Laboratory control samples were obtained from U.S. EPA Environmental monitoring and Support Laboratory, Cincinnati, and were prepared as directed. Results are presented in Table A-2. Precision goals for total organic carbon and total organic halide determinations (both liquid and solid) were met. However, the success in meeting the trace metals precision goals cannot be determined since duplicate analyses as opposed to triplicate analyses were performed. All accuracy and completeness goals were met, except for the total organic halides accuracy analyses on the solids which was not performed in a deviation from the QA plan.

Facility C

Quality control procedures for solids, extractable and volatile organics determination included the analysis of duplicate aliquots of sample and matrix spiked samples by GC/FID. Samples were directly injected so no method blank was prepared. A field bias blank, collected with the samples was analyzed and found to contain less than 0.1 percent Freon TF and 1,1,1-trichloroethane. Laboratory control samples were not available. Precision, accuracy, and completeness goals were met as indicated in Table A-3 except accuracy goals for total solids which were not set in the Facility C QA project plan.

Facility D

Quality control procedures for volatile organics determinations by GC/FID included the preparation and analysis of a laboratory method blank, by which final results were corrected, a field bias blank, duplicate sample injections and matrix spikes of duplicate sample aliquots. A field bias blank was collected along with the samples to measure possible contamination from

TABLE A-1.   FACILITY A QUALITY ASSURANCE SUMMARY

Parameter	Precision		Accuracy		Completeness	
	QA objective (% RPD)[a]	Com-pliance (%)	QA objective (% recovery)	Com-pliance (%)	QA objective (%)[b]	Com-pliance (%)
Trace Metals[c]	≤30	100	70-130	100	95	100
Trace Metals[d]	≤50	100	50-150	100	95	100
Total Organic[c] Carbon	≤30	100	70-130	100	95	100
Total Organic[d] Carbon	≤50	100	50-150	100	95	100
Total Organic[e] Halide	≤30	100	70-130	25	95	100
Total Cyanides	≤30	f	70-130	18	95	0
Total Cyanides	≤50	f	50-150	100	95	0
Hexavalent Chrome	≤30	100	70-130	9	95	0
Hexavalent Chrome	≤50	g	50-150	100	95	0

[a]RPD = Relative Percent Difference.

[b]Percentage of all measurements whose results are judged valid.

[c]Liquid waste matrix.

[d]Solid waste matrix.

[e]Accuracy and precision analyses for solid wastes matrix were not performed.

[f]Interfering substance (excess chlorine) rendered precision analysis results invalid.

[g]Accuracy and precision analyses for solid wastes matrix were not performed.

TABLE A-2. FACILITY B QUALITY ASSURANCE SUMMARY

Parameter	Precision		Accuracy		Completeness	
	QA objective (% RPD)[a]	Com- pliance (%)	QA objective (% recovery)	Com- pliance (%)	QA objective (%)[b]	Com- pliance (%)
Copper	≤ 30	100	70–130	100	95	100
Nickel	≤ 30	100	70–130	100	95	100
Lead	≤ 30	0[c]	70–130	100	95	50
Zinc	≤ 30	100	70–130	100	95	100
Total Organic Carbon	≤ 30[d]	100	70–130	100	95	100
Total Organic Carbon	≤ 50[e]	100	50–150	100	95	100
Total Organic Halide	≤ 30[d]	100	70–130	100	95	100
Total Organic Halide	≤ 50[e]	100	50–150	-f	95	100

[a]RPD = Relative Percent Difference.

[b]Percentage of all measurements whose results are judged valid.

[c]Duplicate analyses instead of triplicate analyses were performed.

[d]Liquid wastes.

[e]Solid wastes.

[f]Accuracy was not measured on solids.

TABLE A-3.   FACILITY C QUALITY ASSURANCE SUMMARY

Parameter	Precision		Accuracy		Completeness	
	QA objective (% RPD)[a]	Com-pliance (%)	QA objective (% recovery)	Com-pliance (%)	QA objective (%)[b]	Com-pliance (%)
Volatile Organics	$\leq$40	100	50-160	100	95	100
Total[c] Solids	$\leq$50	100	-	-	-	-
Extractable Organics	$\leq$75	100	10-150	100	95	100

[a]RPD = Relative Percent Difference.

[b]Percentage of all measurements whose results are judged valid.

[c]Precision and accuracy goals for total solids were not set in the Facility C Quality Assurance Plan.

handling and storage. Quality control procedures for volatile and extractable organic compounds determinations by GC/MS at ERT Analytical Laboratory included the analysis of a method blank, and surrogate and matrix spikes from duplicate sample aliquots. Percent recovery of matrix spiked compounds was calculated as a measure of analytical accuracy. Results of these analyses, presented in Table A-4, indicate that precision and accuracy goals were met and completeness, defined as the percentage of all measurements whose results are judged valid, was determined to be 100 percent.

Quality control procedures for solids determination included the preparation and analysis of a laboratory method blank, by which final results were corrected, and duplicate sample aliquots. Laboratory control samples and matrix spiked samples were not available for analysis because of the nature of the sample matrix, therefore accuracy cannot be determined. Precision goals, set at ≤50 relative percent difference, were met. Precision of analysis conducted on water matrices was not determined.

Facility E

Standard Quality control procedures were implemented whenever possible for program analysis including analysis of a laboratory method blank, an LCS, duplicate sample aliquots, and a matrix spike of duplicate sample aliquots. Completeness for all analyses was 100 percent. Trace metals precision results (≤30 RPD) as indicated in Table A-5 were not met for tin and lead, while accuracy goals (75-125 percent recovery) were not met for tin. Accuracy goals for total organic carbon were not met while precision and completeness goals were. Quality control procedures for volatile organics indicate that both precision (≤50 RPD) and accuracy (50 to 160 percent recovery) goals were met. However, matrix interferences during sample extraction and subsequent sample dilutions reduced spike concentrations on extractable organics to below detection limits (5 µg/L).

Facility F

Quality control procedures for trace metals determination included the analysis of a method blank, by which final results were corrected, a laboratory control sample (LCS), duplicate sample aliquots and matrix spikes of duplicate sample aliquots. LCS's were provided by U.S. EPA Environmental Monitoring and Support Laboratory, Cincinnati. Due to the complexity of the sample matrix, precision and accuracy goals for trace metals were difficult to meet as indicated in Table A-6. Completeness was determined to be 0 percent for tin, 91 percent for lead and 100 percent for copper. All tin results were considered invalid. Lead results on the plating bath were also considered invalid. Precision goals for total organic carbon, set at ≤20 relative percent difference, were met. Matrix spiked samples were prepared by spiking a duplicate sample aliquot with known concentrations of the compounds of interest. Results, provided in Table 61, indicate that accuracy goals (80 to 120 percent recovery) were met. Completeness for total organic carbon was 100 percent.

TABLE A-4.   FACILITY D QUALITY ASSURANCE SUMMARY

Parameter	Precision		Accuracy		Completeness	
	QA objective (% RPD)[a]	Compliance (%)	QA objective (% recovery)	Compliance (%)	QA objective (%)[b]	Compliance (%)
Total Solids	≤ 50	100	---[c]		---[c]	
Volatile Organics	≤ 40	100	50-160	100	95	100
Volatile Organics	≤ 75	100	50-160	100	95	100
Extractable Organics	≤ 75	100	10-150	100	95	100

[a]RPD = Relative Percent Difference.

[b]Percentage of all measurements whose results are judged valid.

[c]Precision and accuracy goals were not set in Facility D QA Plan.

[d]GC/FID Analysis.

[e]GC/MS Analysis (ERT).

TABLE A-5.   FACILITY E QUALITY ASSURANCE SUMMARY

Parameter	Precision		Accuracy		Completeness	
	QA objective (% RPD)[a]	Compliance (%)	QA objective (% recovery)	Compliance (%)	QA objective (%)[b]	Compliance (%)
Tin	≤ 30	50	75-125	50	95	100
Lead	≤ 30	50	75-125	100	95	100
Copper	≤ 30	100	75-125	100	95	100
Total Organic Carbon	≤ 30	100	75-125	50	95	100
Volatile Organics	≤ 50	100	50-160	100	95	100
Extractable Organics	---[c]		---		---	

[a]RPD = Relative Percent Difference
[b]Percentage of all measurements whose results are judged valid
[c]Matrix interference rendered analyses inconclusive

TABLE A-6.   FACILITY F QUALITY ASSURANCE SUMMARY

Parameter	Precision		Accuracy		Completeness	
	QA objective (% RPD)[a]	Compliance (%)	QA objective (% recovery)	Compliance (%)	QA objective (%)[b]	Compliance (%)
Copper	≤ 20	100	80-120	100	95	100
Tin	≤ 20	0	80-120	0	95	0
Lead	≤ 20	100	80-120	90	95	91
Total Organic Carbon	≤ 20	100	80-120	100	95	100

[a]RPD = Relative Percent Difference
[b]Percentage of all measurements whose results are judged valid

SAMPLING PROCEDURES

The sampling procedures outlined in Section 4 of the QA Plans were followed with minor deviations. These sampling procedure deviations have been presented in detail in the individual draft final reports. However, to preserve clarity in the summary report only major sampling procedure changes for each facility will be addressed.

Facility A

- Volatile organic analysis were eliminated from the program as they were noncritical parameters in the metals reduction evaluation and added substantially to the program cost.

- Since it was difficult to sample one batch completely due to length of time necessary to fill the SBH filter press, as much data from a single batch (85-12-1009) was collected as possible. In telephone conversations with the EPA Project Officer, it was agreed that data from separate batches would be acceptable.

Facility B

- A flow meter malfunction necessitated the use of flowrate estimates. The estimates were obtained by contacting the Orange County Sewer Authority for recent data on Facility B wastewater flowrates, thus verifying this data by calculating the throughput of the SBH/ultrafiltration wastewater feed pump.

Facility C

- Samples for TOX (total organic halide) were not collected. The high corrosivity of some of the samples may have adversely affected the analytical instruments.

Facility D

- The total metals analyses proposed were not conducted in order to reduce program analytical costs in accordance with the revised proposal to EPA Project Monitor Harry Freeman dated 28 February 1986.

Facility E

- The electrolytic recovery system was not tested for two reasons: 1) Facility E had inadvertently dumped the static rinse batch which GCA had planned to sample; and 2) delays at Facility E made testing of the new system difficult under the time frame for this program.

Facility F

- Sampling of process and waste streams associated with the box distillation process were eliminated. This was done at the request of the EPA project officer to cut costs.

SAMPLE CUSTODY

Sample custody procedures described in Section 5 of the QA Plans were followed during the sampling program.

CALIBRATION PROCEDURES AND FREQUENCY

Calibration procedures described in Section 6 of the QA Plans were followed during the sampling program.

ANALYTICAL PROCEDURES

Analytical procedures described in Section 7 of the QA Plans were followed during the sampling program.

DATA REDUCTION, VALIDATION, AND REPORTING

Data reduction, validation, and reporting procedures described in Section 8 of the QA Plans were followed during this program.

INTERNAL QUALITY CONTROL CHECKS

Internal QC procedures described in Section 9 of the QA Plans were followed during this program.

PREVENTIVE MAINTENANCE

Preventive maintenance procedures described in Section 11 of the QA Plans were followed during this program.

ASSESSMENT OF PRECISION, ACCURACY AND COMPLETENESS

Analytical precision was reported in terms of relative percent difference using the following equation:

$$RPD = \frac{X_1 - X_2}{\bar{X}} \times 100$$

where: RPD = relative percent difference

X_1 = larger individual measurement

X_2 = smaller individual measurement

$\bar{X}$ = average of X_1 and X_2

Accuracy assessments were based on the results of analyses of EPA Standard Reference Materials and of matrix spiked samples and reported in terms of percent recovery which was calculated as shown below:

$$\text{Percent Recovery} = 100\left(\frac{\text{Measured Value}}{\text{True Value}}\right)$$

The following formula was used to estimate completeness:

$$C = 100\left(\frac{V}{T}\right)$$

C = Percent completeness
V = Number of measurement judged valid
T = Total number of measurements

CORRECTIVE ACTION

There were no Corrective Action Request forms initiated in regard to this program.

QUALITY ASSURANCE REPORTS

All pertinent quality control data and activities have been summarized in this Final Report.

HANDBOOK OF CONTAMINATION CONTROL IN MICROELECTRONICS
Principles, Applications and Technology

Edited by

Donald L. Tolliver
Motorola, Inc.

Contamination control technology is now a prerequisite of modern electronics. This has not always been the case. However, since about 1980, advanced microelectronic circuitry has increased dramatically in its complexity and degree of integration or density of active components, thus necessitating meticulous contamination control. The one megabit DRAM is in production; the 4 megabit device is forecast for 1990; 16 and 64 megabit capacities are in the planning stages; and the Japanese envision a 100 megabit device. Obviously, device defect density is or will be so critical to the successful manufacturing of these devices that only the most astute companies with advanced contamination control technology will be able to survive in the marketplace. For this very basic reason, this handbook will have a timely and important role to play in the industrial marketplace.

This book introduces contamination control in a relatively comprehensive manner. It covers the basics in most areas for the beginner, and it delves in depth into the more critical issues of process engineering and circuit manufacturing for the more advanced reader. The reader will begin to see how the puzzle of contamination control comes together and to focus on the fundamentals required for excellence in modern semiconductor manufacturing.

What makes the arena of contamination control unique is its ubiquitous nature, across all facets of semiconductor manufacturing. Clean room technology, well recognized as a fundamental requirement in modern day circuit manufacturing, barely scratches the surface in total contamination control. This handbook makes the first attempt to define and describe most of the major categories in current contamination control technology.

CONTENTS

ISBN 0-8155-1151-5 (1988)

488 pages

Printed and bound by CPI Group (UK) Ltd, Croydon, CR0 4YY

03/10/2024

01040335-0016